Christian Koos

Nanophotonic Devices for Linear and Nonlinear Optical Signal Processing

Karlsruhe Series in Photonics & Communications, Vol. 1
Edited by Prof. J. Leuthold and Prof. W. Freude

Universität Karlsruhe (TH), Institute of High-Frequency and Quantum Electronics (IHQ),
Germany

Nanophotonic Devices for Linear and Nonlinear Optical Signal Processing

von
Christian Koos

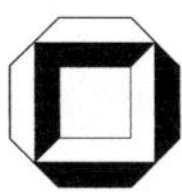

universitätsverlag karlsruhe

Dissertation, Universität Karlsruhe (TH)
Fakultät für Elektrotechnik und Informationstechnik, 2007

Impressum

Universitätsverlag Karlsruhe
c/o Universitätsbibliothek
Straße am Forum 2
D-76131 Karlsruhe
www.uvka.de

Universitätsverlag Karlsruhe 2007
Print on Demand

ISSN: 1865-1100
ISBN: 978-3-86644-178-1

Nanophotonic Devices for Linear and Nonlinear Optical Signal Processing

Zur Erlangung des akademischen Grades eines

DOKTOR-INGENIEURS

der Fakultät für
Elektrotechnik und Informationstechnik
der Universität Fridericiana Karlsruhe

genehmigte

DISSERTATION

von

Dipl.-Ing. Christian Koos

aus

Heilbronn

Tag der mündlichen Prüfung:	10. Juli 2007
Hauptreferent:	Prof. Dr. Dr. h. c. Wolfgang Freude
Korreferenten:	Prof. Dr. Jürg Leuthold
	Prof. Dr. Michael Siegel

To my parents

So eine Arbeit wird eigentlich nie fertig,
man muß sie für fertig erklären, wenn man nach Zeit
und Umständen das Möglichste getan hat.

J. W. Goethe, Italienische Reise,
Caserta, den 16. März 1787

Contents

List of Figures

List of Tables

Abstract (German)

Die vorliegende Arbeit befasst sich mit nanophotonischen Bauteilen zur linearen und nichtlinearen optischen Signalverarbeitung. Es werden Aspekte der Modellierung, Herstellung und Charakterisierung solcher Bauteile diskutiert. Der engen Verflechtung dieser Schritte wird durch eine ganzheitliche Betrachtungsweise Rechnung getragen. Darüber hinaus werden neuartige Bauteilkonzepte entwickelt und ein spezielles technologisches Verfahren zu ihrer Herstellung untersucht.

Nanophotonische Wellenleiter müssen einen hohen Brechungsindexunterschied zwischen Kern- und Mantelbereich aufweisen, um die Abstrahlverluste auch bei sehr kleinen Kurvenradien (typischerweise $< 10\,\mu$m) gering zu halten. Während für herkömmliche Wellenleiter mit kleinem Indexkontrast eine skalare Annäherung der Modenfelder befriedigende Ergebnisse liefert, wird zur Modellierung von Strukturen mit hohem Indexkontrast eine vektorielle Beschreibungsweise benötigt. Daher wird in der vorliegenden Arbeit zunächst die lineare und nichtlineare Kopplung von Wellenleitermoden unter Berücksichtigung vektorieller Modenfelder untersucht.

Bei Wellenleitern mit hohem Indexkontrast können bereits geringe Seitenwandrauhigkeiten zu starken Streuverlusten führen. Auf Grundlage der vektoriellen Modenkopplungstheorie wird eine neuartige Methode zur quantitativen Beschreibung von Streuverlusten vorgestellt. Die Ergebnisse werden durch numerische Simulationen und durch Experimente bestätigt, und es werden Richtlinien für das Design besonders verlustarmer gerader Wellenleiter entwickelt.

Für gekrümmte Wellenleiter wird durch einen Variationsansatz eine Klasse idealer Konturkurven ermittelt, die niedrige Streuverluste aufweisen und gleichzeitig tolerant gegenüber Lithographieverfahren mit begrenzter Auflösung sind. Die Ergebnisse werden durch numerische Simulationen und Experimente bestätigt.

Manche Anwendungen wie z.B. elektro-optische Modulatoren beruhen auf geometrisch sehr einfachen Wellenleitern, die jedoch sehr glatte Seitenwände aufweisen müssen. Für solche Wellenleiter wurde im Rahmen der vorliegenden Arbeit ein spezielles Herstellungsverfahren entwickelt. Es beruht auf der kristallographisch-anisotropen Ätzung von einkristallinen Ausgangsmaterialien und erlaubt die Strukturierung von Wellenleitern mit atomar glatten Seitenwänden. Das Herstellungsverfahren wird im Detail beschrieben, und es werden Prototypen vorgestellt.

Breitbandige, auf Silicon-on-Insulator-(SOI-)Substraten integrierbare elektro-optische Modulatoren sind von beträchtlichem technischem wie kommerziellem Interesse, insbesondere vor dem Hintergrund, dass optische Übertragungsverfahren zunehmend für die Kommunikation über mittlere (rack-to-rack) und kurze (board-to-board, chip-to-chip)

Entfernungen eingesetzt werden. In dieser Arbeit wird ein neuartiges Konzept für kompakte, breitbandige SOI-Modulatoren entwickelt und untersucht. Die erzielbaren Bandbreiten ermöglichen Datenraten von über 100 Gbit/s.

Kerr-Nichtlinearitäten ermöglichen eine schnelle, voll-optische Signalerverarbeitung, beispielsweise die Regeneration oder Wellenlängenkonversion von Nachrichtensignalen. Hierbei bieten sich nanophotonische Wellenleiter aufgrund der starken Feldkonzentration an. Besondere Freiheitsgrade im Design erhält man, wenn die Wechselwirkung des im Wellenleiterkern geführten Lichtes mit einem nichtlinearen Mantelmaterial ausgenutzt wird. In der vorliegenden Arbeit werden nichtlineare Streifen- und Schlitzwellenleiter mit unterschiedlichen Mantelmaterialien untersucht und auf höchstmögliche Nichtlineariät optimiert. Für die veschiedenen Wellenleiterstrukturen werden Designkurven vorgestellt und die zugehörigen nichtlinearen Wellenleiterparameter γ abgeschätzt. Dabei werden Werte von bis zu $\gamma = 7 \times 10^3 \, (\mathrm{W\,m})^{-1}$ erreicht, die mehr als drei Größenordnungen über den Werten für modernste hoch-nichtlineare Glasfasern liegen.

Werden optische Halbleiterverstärker als nichtlineare Bauelemente zur optischen Signalverarbeitung verwendet, so müssen die zugehörigen Ladungsträgerdynamiken berücksichtigt werden. Typische Zeitskalen liegen dabei im einstelligen Pikosekundenbereich. Die Ladungsträgerdynamik beeinflusst sowohl Amplitude als auch Phase des Ausgangssignals. Im Rahmen der vorliegenden Arbeit wurde in Pump-Probe-Experimenten die komplexwertige zeitaufgelöste Übertragungsfunktion eines InAs/GaAs-Quantenpunktverstärkers bei einer Wellenlänge von $\lambda = 1{,}3 \, \mu\mathrm{m}$ gemessen. Die Antwortzeiten betragen typischerweise 3 ps, wobei sich die Amplitude des (kleinen) Probe-Signals stark, die Phase aber nur verhältnismäßig schwach ändert. Eine effiziente optische Signalverabeitung in solchen Bauteilen sollte also auf Basis von Kreuzgewinnmodulation (Cross-gain modulation, XGM) und nicht durch Kreuzphasenmodulation (Cross-phase modulation, XPM) erfolgen.

Die für nichtlineare optische Wechselwirkungen benötigten Pumpleistungen können verringert werden, indem die Feldüberhöhung in resonanten Strukturen gezielt ausgenutzt wird. Dies wird am Beispiel eines nichtlinearen Ringresonators experimentell demonstriert. Darauf aufbauend werden verschiedene Finite-Differenzen-Methoden (engl. finite-difference time-domain (FDTD) method) hinsichtlich ihres Rechenaufwandes und ihrer Zuverlässigkeit untersucht. Es stellt sich heraus, dass die zweidimensionale Berechnung einer planaren Struktur mit einem verfeinerten FDTD-Algorithmus weitaus bessere Ergebnisse liefert als die numerisch aufwendigere dreidimensionale Modellierung mit einem Standard-FDTD-Algorithmus.

Preface

Irrespective of telecommunication market cycles, data traffic has vastly increased in the last years. The development of broadband access networks has contributed significantly to this growth: Slow dial-up connections have been widely replaced by digital subscriber line (xDSL) techniques, which are currently being substituted by optical systems (fiber to the home, FTTH). Providing additional broadband services (e.g., Video on Demand) over these networks will foreseeably push the data rates of the current backbone infrastructure to the limits, and new technologies will become necessary. The Institute of Electrical and Electronics Engineers (IEEE) is currently working on the next IEEE 802 standard for 100 Gbit/s-Ethernet, and major vendors have as recently as of March 2007 made predictions that next-generation networks will be based on optical 100 Gbit/s-Ethernet or 100 Gbit/s Optical Transport Network (OTN, ITU-T G.709). On a medium-term perspective, optical transmission techniques are expected to replace electrical interconnects on the rack-to-rack, board-to-board, and even chip-to-chip level.

Photonic integrated circuits (PIC) play a major role in this development: First, the high-volume deployment of optical interconnects requires compact passive optical devices and electro-optical modulators that can be cheaply mass-produced. Second, the constantly increasing data rates in backbone networks require novel high-performance devices for ultrafast all-optical signal processing.

Integrated nanophotonic circuits meet these demands to an excellent extent: Highly parallel microfabrication techniques lend themselves for cheap large-scale production. The strong guidance of light allows for strong field confinement and therefore enables dense integration and exploitation of nonlinear effects at low power. Particularly intense research is currently carried out to integrate optical devices on silicon substrates, using mature complementary metal-oxide-semiconductor (CMOS) fabrication techniques [13, 67, 111]. To achieve PIC with the sophistication of electrical integrated circuits, robust techniques for design, fabrication and packaging have to be developed.

The scope of this thesis is the conception, design, modelling, fabrication and characterization of integrated nanophotonic devices. Since there is strong interaction between these steps, special emphasis is placed on considering them as an entity.

The thesis is structured as follows: In Chapter 1, the fundamental electromagnetic theory of integrated nanophotonic waveguides is introduced, and coupled-mode equations are derived for the linear and nonlinear case. Mathematical definitions and derivations are shifted to the Appendices A, B, and C. Publications and Patents are summarized on pp. 199.

In Chapter 2, we describe the experimental setups and the measurement techniques that are needed for the linear and nonlinear characterization nanophotonic devices. These techniques are crucial for the experimental results of the subsequent chapters.

Optical waveguides in HIC material systems are prone to roughness-induced scattering loss. In Chapter 3, the scattering mechanism is investigated both theoretically and experimentally. Mathematical details on the calculation of radiation mode fields are discussed in Appendix D. Having understood the origin of waveguide loss, we can give design guidelines for low-loss straight waveguides. The findings have been published in a journal article (J2) and several international conference contributions (C17, C10, C9, C8, C7, C6, C4).

For curved sections, losses can be reduced by using appropriate contour trajectories. In Chapter 4, a class of contour trajectories for minimum radiation loss is derived analytically, and numerical optimization is used to find ideal trajectory parameters. The contour trajectories exhibit maximum robustness with respect to finite lithographic resolution. 3D FDTD simulations predict bend losses below 0.1dB for 180°-bends of overmoded silicon-on-insulator (SOI) waveguides with a radius of $R = 1.5\,\mu\text{m}$. These findings are supported experimentally. The results have been published in a journal article (J4) and several international conference contributions (C17, C15, C13).

For special applications, reducing the scattering loss by design is not possible. For these cases, we develop a novel method to fabricated waveguide structures with atomically smooth surfaces. This method is presented in Chapter 5. It is based on preferential wet etching of crystalline materials and is particularly well suited for optical waveguides in the single-crystalline device layer of silicon-on-insulator (SOI) wafers. A patent has been filed for the fabrication method (P2). Publication is envisaged once the patent is accepted.

Second-order nonlinear interaction is studied in Chapter 6, where we propose and investigate novel SOI modulator schemes with unprecedented electrical bandwidth and compactness. Our design enables 100 Gbit/s modulation at 3 V peak-to-peak voltage with compact devices of less than 2 mm length. A patent has been filed on this invention (P1), and publications are being prepared for submission once the patent is accepted.

Third-order nonlinear SOI strip and slot waveguides for all-optical signal processing are investigated in Chapter 7. At $\lambda = 1.55\,\mu\text{m}$ we provide universal design curves for strip and slot waveguides which are covered with different linear and nonlinear materials, and we calculate the resulting maximum nonlinear waveguide parameter γ. If properly designed, values up to $\gamma = 7 \times 10^3\,(\text{W\,m})^{-1}$ can be achieved. This is more than three orders of magnitude larger than for state-of-the-art highly nonlinear fibers. The instantaneous response of Kerr-type nonlinearities enables ultrafast all-optical signal processing with nonresonant compact devices. The results have been published in a journal article (J5) and in international conference contributions (C18, C19).

In contrast to devices based on Kerr nonlinearities, nonlinear semiconductor optical amplifiers (SOA) exhibit carrier dynamics with typical response times in the low ps-range. To design SOA-based all-optical switches or wavelength converters, the magnitudes and dynamics of gain and phase effects must be known. Chapter 8 describes pump-probe characterization of nonlinear dynamics of an InAs/GaAs quantum dot (QD) amplifier at 1300 nm. Strong 3 ps gain variations with only weak phase changes are measured. Such low-alpha factor devices are well suited for cross-gain modulation based signal processing.

The results have been published in an international conference contribution (C16). A journal article is being prepared.

Nonlinear interaction can be significantly enhanced when using resonant structures. For complex waveguide geometries, however, a quantitatively reliable analysis requires computationally expensive numerical modelling. It is therefore of high interest to know which degree of complexity is needed in order to obtain reliable simulation results. In Chapter 9, we experimentally investigate resonant enhancement of nonlinear optical interaction in a ring resonator, and we study the accuracy of different nonlinear finite-difference time-domain (FDTD) models by comparing the results of a series of progressively less complex implementations. Incorporating staircasing error correction and material dispersion into a two-dimensional effective index model turns out to be computationally much cheaper and more effective than performing a fully three-dimensional simulation without these features. The results have been published in journal articles (J3, J1) and in international conference contributions (C14, C12, C11).

Chapter 10 finally summarizes the work and gives and an outlook on future research.

Chapter 1

Theoretical Background

In the following we introduce the basic electromagnetic theory that is needed to describe both linear and nonlinear optical phenomena in integrated optical waveguides. Parts of the formalisms can be considered standard knowledge and are subject of various textbooks. A particularly comprehensive presentation of linear waveguide theory is developed in [72], [74], and [81]. Nonlinear optical phenomena in bulk materials are extensively studied in [14], and [95] considers nonlinearities in low index-contrast waveguides. [3] is the standard reference for nonlinear effects in optical fibers.

However, existing formalisms are adapted to low index-contrast material systems, and it is usually assumed that $\nabla \cdot \mathbf{E} = \epsilon^{-1} \nabla \cdot \mathbf{D}$, which requires $\nabla \left(\epsilon^{-1} \right) \approx 0$ in the entire cross section of the waveguide (see for example Eq. (2.1.18) in [3]). The whole analysis can then be formulated in terms of the scalar electric mode fields only. This approximation is excellent for optical fibers and other low index-contrast systems, but it does not hold for integrated high index-contrast (HIC) structures, and the accuracy of standard equations for fibers is at least questionable when applied to HIC waveguides. We therefore derive a mode coupling formalism which uses vectorial mode fields rather than scalar approximations and which can be applied to linear and nonlinear HIC index-contrast waveguides.

This chapter is structured as follows: In Section 1, electromagnetic basics are reviewed, including time domain and frequency domain formulations of Maxwell's equations with nonlinear polarization. In Section 2, the concept of waveguide modes is introduced and signal propagation in ideal linear waveguides is studied. Linear coupling of waveguide modes is discussed in Section 3. Sections 4 and 5 are dedicated to second- and third-order nonlinear interaction in HIC waveguides.

1.1 Electromagnetic Basics

1.1.1 Maxwell's Equations and Nonlinear Polarization

In the absence of any free carriers and currents, Maxwell's equations take the following form [49]:

$$\nabla \times \mathbf{H}(\mathbf{r}, t) = \frac{\partial \mathbf{D}(\mathbf{r}, t)}{\partial t} \tag{1.1}$$

$$\nabla \times \mathbf{E}(\mathbf{r}, t) = -\frac{\partial \mathbf{B}(\mathbf{r}, t)}{\partial t} \tag{1.2}$$

$$\nabla \cdot \mathbf{D}(\mathbf{r}, t) = 0 \tag{1.3}$$

$$\nabla \cdot \mathbf{B}(\mathbf{r}, t) = 0 \tag{1.4}$$

The vector $\mathbf{r} = (x, y, z)^{\mathrm{T}}$ defines a point in three-dimensional space. For optical waveguides, the medium is assumed to be nonmagnetic. The magnetic flux density $\mathbf{B}$ is then related to the magnetic field $\mathbf{H}$ by

$$\mathbf{B}(\mathbf{r}, t) = \mu_0 \mathbf{H}(\mathbf{r}, t), \tag{1.5}$$

where $\mu_0 = 1.25664 \times 10^{-6}\,\mathrm{V\,s/(A\,m)}$ is the magnetic permeability of vacuum. The relation between the electric field $\mathbf{E}$ and the electric displacement $\mathbf{D}$ can be expressed as

$$\mathbf{D}(\mathbf{r}, t) = \epsilon_0 \mathbf{E}(\mathbf{r}, t) + \mathbf{P}(\mathbf{r}, t), \tag{1.6}$$

where $\epsilon_0 = 8.85419 \times 10^{-6}\,\mathrm{A\,s/(V\,m)}$ is the electric permittivity of vacuum, and where $\mathbf{P}$ denotes the polarization.

Assuming a dielectric material which is local in space, the polarization $\mathbf{P}$ depends only upon the local value of the electric field $\mathbf{E}$. In the most general case, the medium's polarization response is nonlinear and nonlocal in time. The relation between $\mathbf{E}$ and $\mathbf{P}$ can be expanded into a Volterra series,

$$\mathbf{P}(\mathbf{r}, t) = \epsilon_0 \int_{-\infty}^{+\infty} \underline{\chi}^{(1)}\,(t - \tau_1)\,\mathbf{E}(\mathbf{r}, \tau_1)\,\mathrm{d}\tau_1 \tag{1.7}$$

$$+ \epsilon_0 \iint_{-\infty}^{+\infty} \underline{\chi}^{(2)}\,(t - \tau_1, t - \tau_2) : \mathbf{E}(\mathbf{r}, \tau_1)\mathbf{E}(\mathbf{r}, \tau_2)\,\mathrm{d}\tau_1\,\mathrm{d}\tau_2$$

$$+ \epsilon_0 \iiint_{-\infty}^{+\infty} \underline{\chi}^{(3)}\,(t - \tau_1, t - \tau_2, t - \tau_3)\;\vdots\;\mathbf{E}(\mathbf{r}, \tau_1)\mathbf{E}(\mathbf{r}, \tau_2)\mathbf{E}(\mathbf{r}, \tau_2)\,\mathrm{d}\tau_1\,\mathrm{d}\tau_2\,\mathrm{d}\tau_3$$

$$+ \ldots$$

Since both the dielectric polarization $\mathbf{P}$ and the electric field $\mathbf{E}$ are vectorial quantities, the time-domain Volterra kernel of n-th order, $\underline{\chi}^{(n)}\,(t_1, t_2, \ldots, t_n)$, is a tensor of rank $n+1$. The tensorial nature of $\underline{\chi}^{(n)}\,(t_1, t_2, \ldots, t_n)$ is discussed in more detail in Appendix B. Since the polarization response of the medium is causal, the components of $\underline{\chi}^{(n)}(t_1, t_2, \ldots, t_n)$ vanish for negative time arguments, see Eq. (B.8). The spatial dependence of $\underline{\chi}^{(n)}(t_1, t_2, \ldots, t_n)$ is omitted for the sake of readability.

The electric polarization $\mathbf{P} = \mathbf{P}^{(\mathrm{lin})} + \mathbf{P}^{(\mathrm{nl})}$ can be split up into a part $\mathbf{P}^{(\mathrm{lin})}$ that depends linearly on $\mathbf{E}$,

$$\mathbf{P}^{(\mathrm{lin})}(\mathbf{r}, t) = \epsilon_0 \int_{-\infty}^{+\infty} \underline{\chi}^{(1)}(t - \tau_1)\, \mathbf{E}(\mathbf{r}, \tau_1)\, \mathrm{d}\tau_1, \tag{1.8}$$

and into the remaining nonlinear contribution $\mathbf{P}^{(\mathrm{nl})}$,

$$\mathbf{P}^{(\mathrm{nl})}(\mathbf{r}, t) = \epsilon_0 \iint_{-\infty}^{+\infty} \underline{\chi}^{(2)}(t - \tau_1, t - \tau_2) : \mathbf{E}(\mathbf{r}, \tau_1)\mathbf{E}(\mathbf{r}, \tau_2)\, \mathrm{d}\tau_1\, \mathrm{d}\tau_2 \tag{1.9}$$

$$+ \epsilon_0 \iiint_{-\infty}^{+\infty} \underline{\chi}^{(3)}(t - \tau_1, t - \tau_2, t - \tau_3)\, \vdots\, \mathbf{E}(\mathbf{r}, \tau_1)\mathbf{E}(\mathbf{r}, \tau_2)\mathbf{E}(\mathbf{r}, \tau_2)\, \mathrm{d}\tau_1\, \mathrm{d}\tau_2\, \mathrm{d}\tau_3$$

$$+ \ldots .$$

For most cases of practical interest, the nonlinear polarization $\mathbf{P}^{(\mathrm{nl})}$ is much weaker than the linear polarization $\mathbf{P}^{(\mathrm{lin})}$. Optical nonlinearities can hence be treated as a small perturbation of the solution for the linear case.

For optical signals of limited bandwidth, the convolution in Eq. (1.8) can be approximated by a multiplication with the linear optical susceptibility tensor $\widetilde{\underline{\chi}}^{(1)}$, see Eq. (B.22). The linear contribution to the dielectric displacement $\mathbf{D}$ can then be expressed by the dielectric permeability tensor $\underline{\epsilon}_r$,

$$\mathbf{D}(\mathbf{r}, t) = \epsilon_0\, \underline{\epsilon}_r(\mathbf{r})\, \mathbf{E}(\mathbf{r}, t), \qquad \underline{\epsilon}_r(\mathbf{r}) = \underline{\mathbf{1}} + \widetilde{\underline{\chi}}^{(1)}, \tag{1.10}$$

where $\underline{\mathbf{I}}$ denotes the unity tensor.

If the medium is isotropic, the first-order susceptibility tensor $\widetilde{\underline{\chi}}^{(1)}$ can be replaced by a scalar $\widetilde{\chi}^{(1)}$. The permeability tensor $\underline{\epsilon}_r$ can then also be replaced by the scalar relative dielectric permeability $\epsilon_r = 1 + \widetilde{\chi}^{(1)}$ or by the refractive index $n = \sqrt{\epsilon_r}$,

$$\mathbf{D} = \epsilon_0 \epsilon_r \mathbf{E} = \epsilon_0 n^2 \mathbf{E}. \tag{1.11}$$

1.1.2 Time- and Frequency-Domain Quantities

For many cases of interest, we can safely neglect the nonlinear contribution to the polarization, $\mathbf{P}^{(\mathrm{nl})} = 0$. Maxwell's equations are then linear, and any dynamic processes can be treated as a superposition of steady-state solutions by means of Fourier transformation,

$$\widetilde{\mathbf{F}}(\mathbf{r}, \omega) = \int_{-\infty}^{+\infty} \mathbf{F}(\mathbf{r}, t)\, e^{-j\omega t}\, dt \tag{1.12}$$

$$\mathbf{F}(\mathbf{r}, t) = \frac{1}{2\pi} \int_{-\infty}^{+\infty} \widetilde{\mathbf{F}}(\mathbf{r}, \omega)\, e^{j\omega t}\, d\omega, \tag{1.13}$$

where the time-domain function $\mathbf{F}(\mathbf{r}, t)$ denotes an electromagnetic field quantity such as $\mathbf{D}$, $\mathbf{E}$, $\mathbf{B}$, $\mathbf{H}$, or $\mathbf{P}$. Since a physically meaningful quantity $\mathbf{F}(\mathbf{r}, t)$ has to be real, its spectrum $\widetilde{\mathbf{F}}(\mathbf{r}, \omega)$ must fulfill the relation

$$\widetilde{\mathbf{F}}(\mathbf{r}, -\omega) = \widetilde{\mathbf{F}}^{*}(\mathbf{r}, \omega). \tag{1.14}$$

The left half of the spectrum $(-\infty < \omega < 0)$ can thus be derived form the right half $(0 \leq \omega < \infty)$. This redundancy is removed by using analytic signals $\underline{\mathbf{F}}(\mathbf{r}, t)$ in the time-domain. The real part of the analytic signal represents the physically relevant quantity $\mathbf{F}(\mathbf{r}, t)$,

$$\mathbf{F}(\mathbf{r}, t) = \mathrm{Re}\left\{\underline{\mathbf{F}}(\mathbf{r}, t)\right\}, \tag{1.15}$$

whereas the imaginary part is chosen such that the Fourier-transform $\widetilde{\underline{\mathbf{F}}}(\mathbf{r}, \omega)$ is single-sided, i.e. $\widetilde{\underline{\mathbf{F}}}(\mathbf{r}, \omega) = 0$ for $\omega < 0$. For this to be the case, the imaginary part of $\underline{\mathbf{F}}(\mathbf{r}, t)$ has is related to the real part by the Hilbert transformation, see Appendix A for more details.

If the spectrum consists of discrete lines at frequencies ω_ν, $\nu = 1 \ldots N$, complex amplitudes $\widehat{\mathbf{F}}$ of harmonic electromagnetic quantities $\mathbf{F}$ are used rather than Fourier transforms,

$$\underline{\mathbf{F}}(\mathbf{r}, t) = \widehat{\mathbf{F}}(\mathbf{r}, 0) + \sum_{\nu=1}^{N} \widehat{\mathbf{F}}(\mathbf{r}, \omega_\nu)\, e^{j\omega_\nu t}, \tag{1.16}$$

$$\mathbf{F}(\mathbf{r}, t) = \widehat{\mathbf{F}}(\mathbf{r}, 0) + \frac{1}{2} \sum_{\nu=1}^{N} \left(\widehat{\mathbf{F}}(\mathbf{r}, \omega_\nu)\, e^{j\omega_\nu t} + \mathrm{cc} \right), \tag{1.17}$$

where "cc" stands for the complex conjugate of the preceding expression. $\underline{\widehat{\mathbf{F}}}(\mathbf{r}, 0) = \widehat{\mathbf{F}}(\mathbf{r}, 0)$ is a real number and represents a DC-component. It is often convenient to extend the sum to zero and negative frequencies by defining $\omega_0 = 0$, $\omega_{-\nu} = -\omega_\nu$, and

$$\widehat{\mathbf{F}}(\mathbf{r}, -\omega_\nu) = \widehat{\mathbf{F}}^{*}(\mathbf{r}, \omega_\nu), \tag{1.18}$$

where the asterisk* indicates the complex conjugate. Eqs. (1.16) and (1.17) can then be written as

$$\underline{\mathbf{F}}(\mathbf{r}, t) = \sum_{\nu=0}^{N} \widehat{\underline{\mathbf{F}}}(\mathbf{r}, \omega_\nu)\, \mathrm{e}^{\mathrm{j}\omega_\nu t}, \tag{1.19}$$

$$\mathbf{F}(\mathbf{r}, t) = \frac{1}{2} \sum_{\nu=-N}^{N} (1 + \delta_{\nu,0})\, \widehat{\underline{\mathbf{F}}}(\mathbf{r}, \omega_\nu)\, \mathrm{e}^{\mathrm{j}\omega_\nu t}. \tag{1.20}$$

$\delta_{\nu,0}$ is the Kronecker delta. It is now straightforward to express the Fourier transform of such a signal in terms of its complex amplitudes,

$$\widetilde{\underline{\mathbf{F}}}(\mathbf{r}, \omega) = 2\pi \sum_{\nu=0}^{N} \widehat{\underline{\mathbf{F}}}(\mathbf{r}, \omega_\nu)\delta(\omega - \omega_\nu), \tag{1.21}$$

$$\widetilde{\mathbf{F}}(\mathbf{r}, \omega) = \pi \sum_{\nu=-N}^{N} (1 + \delta_{\nu,0})\, \widehat{\underline{\mathbf{F}}}(\mathbf{r}, \omega_\nu)\delta(\omega - \omega_\nu), \tag{1.22}$$

1.1.3 Energy Flow and Poynting Vector

An important quantity is the energy flow in an electromagnetic field. It is represented by the Poynting vector

$$\mathbf{S}(\mathbf{r},t) = \mathbf{E}(\mathbf{r},t) \times \mathbf{H}(\mathbf{r}, t). \tag{1.23}$$

$\mathbf{S}(\mathbf{r},t)$ is time dependent and represents the instantaneous energy flux. However, in many practical cases, the signals $\mathbf{E}(\mathbf{r},t)$ and $\mathbf{H}(\mathbf{r}, t)$ are harmonics at frequency ω_c and it is rather the average energy flux that is of interest. Defining the complex Poynting vector as the cross product of the corresponding complex amplitudes,

$$\underline{\mathbf{S}}(\mathbf{r}, \omega_\mathrm{c}) = \frac{1}{2} \left(\widehat{\underline{\mathbf{E}}}(\mathbf{r}, \omega_\mathrm{c}) \times \widehat{\underline{\mathbf{H}}}^*(\mathbf{r}, \omega_\mathrm{c}) \right), \tag{1.24}$$

the average energy flux can be obtained by taking the real part of $\underline{\mathbf{S}}$. The net power flow P through an area A is then given by

$$P = \iint_A \mathrm{Re}\,\{\underline{\mathbf{S}}\} \cdot \mathbf{n}\ \mathrm{d}a, \tag{1.25}$$

where $\mathrm{d}a$ is the surface element and $\mathbf{n}$ is the surface normal pointing into the direction in which the energy flow is counted positive.

1.1.4 Slowly-Varying Envelope Approximations

If we take into account the nonlinear polarization $\mathbf{P}^{(\mathrm{nl})}$, the system of Maxwell's equations becomes nonlinear, and the time-domain dynamics can no longer be treated as a direct superposition of frequency-domain solutions. However, as the nonlinear polarization is usually much weaker than the linear contribution, i.e., $\left|\mathbf{P}^{(\mathrm{nl})}\right| << \left|\mathbf{P}^{(\mathrm{lin})}\right|$, its effect may be treated as a perturbation of the solution associated with the linear system by introducing an electric displacement of the form

$$\underline{\mathbf{D}} = \epsilon_0 \epsilon_r \underline{\mathbf{E}} + \underline{\mathbf{P}}^{(\text{nl})}. \tag{1.26}$$

The nonlinear system dynamics are then analyzed by means of a slowly-varying envelope approximation. For this purpose, a "weak" time dependence is assigned to the field amplitude of each carrier frequency ω_ν, and $\underline{\widehat{\mathbf{F}}}(\mathbf{r}, \omega_\nu)$ is replaced by $\underline{\widehat{\mathbf{F}}}(\mathbf{r}, t, \omega_\nu)$ in Eq. (1.19). For nonzero carrier frequency ω_ν, $\underline{\widehat{\mathbf{F}}}(\mathbf{r}, t, \omega_\nu)$ varies "considerably slower than the carrier" both in time and space[1],

$$\underline{\mathbf{F}}(\mathbf{r}, t) = \sum_{\nu=0}^{N} \underline{\widehat{\mathbf{F}}}(\mathbf{r}, t, \omega_\nu)\, \mathrm{e}^{\mathrm{j}\,\omega_\nu t}. \tag{1.27}$$

For DC-components, we again use $\omega_0 = 0$, and the corresponding field $\underline{\widehat{\mathbf{F}}}(\mathbf{r}, t, 0)$ is real. It represents a slowly-varying unmodulated electromagnetic field quantity[2]. If we follow Eq. (1.18) and define

$$\underline{\widehat{\mathbf{F}}}(\mathbf{r}, t, -\omega_\nu) = \underline{\widehat{\mathbf{F}}}^{*}(\mathbf{r}, t, \omega_\nu) \tag{1.28}$$

for $\nu \neq 0$, the real time-domain signal can be expressed as

$$\mathbf{F}(\mathbf{r}, t) = \frac{1}{2} \sum_{\nu=-N}^{N} (1 + \delta_{\nu,0})\, \underline{\widehat{\mathbf{F}}}(\mathbf{r}, t, \omega_\nu)\, \mathrm{e}^{\mathrm{j}\,\omega_\nu t}. \tag{1.29}$$

1.2   Optical Waveguides and Coupling of Modes

An ideal optical waveguide is an arrangement of dielectric material that is uniform in one direction and is capable of guiding light along its axis. It is represented by a refractive index profile $n(\mathbf{r})$ that does not depend on the space coordinate which is parallel to the waveguide axis and which shall be associated with the z-direction,

$$n(x, y, z) = n(x, y, 0). \tag{1.30}$$

In the following, x corresponds to the horizontal direction (parallel to the substrate of an integrated optical waveguide) and the y-axis points in the vertical direction (perpendicular to the plane of the substrate), see Fig. 1.1.

Real waveguides can deviate from this ideal situation in various ways: They may exhibit surface roughness due to an imperfect fabrication process, or any of the core or cladding materials can feature wanted or unwanted optical nonlinearities. In most cases, these deviations can be considered as a space-dependent perturbation of the electric polarization $\mathbf{P}$. This perturbation can usually be assumed to be "weak" in the sense that it only affects a small fraction of the total volume and/or that the relative magnitude of

---

[1] The somewhat fuzzy formulation "considerably slower than the carrier" becomes clear when considering the signal in the frequency domain: The spectra $\underline{\widehat{\mathbf{F}}}(\mathbf{r}, \omega, \omega_\nu)$ of the envelope functions $\underline{\widehat{\mathbf{F}}}(\mathbf{r}, t, \omega_\nu)$ must be narrow enough to prevent spectral overlapping of the envelopes of neighbouring carriers in the expansion $\underline{\widetilde{\mathbf{F}}}(\mathbf{r}, \omega) = \sum_{\nu=0}^{N} \underline{\widehat{\mathbf{F}}}(\mathbf{r}, \omega - \omega_\nu, \omega_\nu)$.

[2] An example of practical interest is the microwave field in an electro-optical modulator.

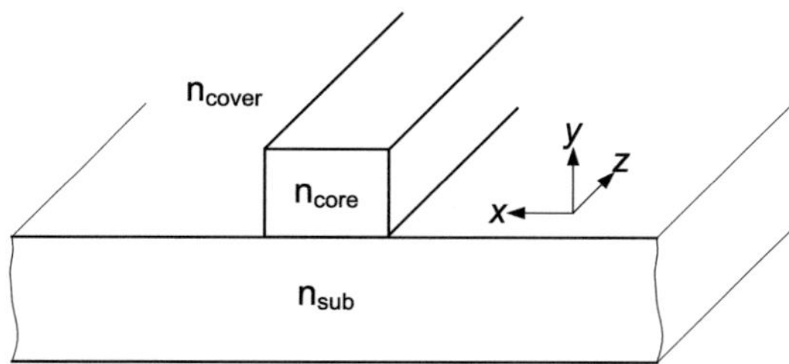

Fig. 1.1. Schematic of an ideal dielectric strip waveguide. The refractive indices of the waveguide core, the substrate and the cover are denoted as n_{core}, n_{sub}, and n_{cover}, respectively.

the perturbation is small. We can then assume that the transverse mode field profiles do not change, and the effects of the perturbation only lead to coupling of waveguide modes. This is the basic principle of coupled-mode theory [72], [74], [81]. A similar approach has been used to describe propagation in nonlinear optical fibers [3], which is however limited to low index-contrast optical waveguides and can therefore not directly be applied to high index-contrast systems.

In the following sections, we develop a consistent treatment of linear and nonlinear coupling of waveguide modes in high index-contrast systems. Details of the derivation can be found in Appendix C.

1.2.1 Modes of Ideal Waveguides

Neglecting any nonlinear optical effects, Maxwell's equations can be treated in the frequency-domain, leading to two coupled differential equations for the electric and the magnetic field,

$$\nabla \times \underline{\widetilde{\mathbf{H}}}(\mathbf{r}, \omega) = j\omega\epsilon_0 n^2(\mathbf{r})\underline{\widetilde{\mathbf{E}}}(\mathbf{r}, \omega) \tag{1.31}$$

$$\nabla \times \underline{\widetilde{\mathbf{E}}}(\mathbf{r}, \omega) = -j\omega\mu_0\underline{\widetilde{\mathbf{H}}}(\mathbf{r}, \omega). \tag{1.32}$$

For a refractive index profile that is uniform in the z-direction, we are particularly interested in solutions of the form

$$\underline{\widetilde{\mathbf{E}}}(\mathbf{r}, \omega) = \mathcal{E}(x, y, \omega)\,e^{-j\beta(\omega)z} \tag{1.33}$$

$$\underline{\widetilde{\mathbf{H}}}(\mathbf{r}, \omega) = \mathcal{H}(x, y, \omega)\,e^{-j\beta(\omega)z}. \tag{1.34}$$

Each of these solutions represents an eigenmode of the waveguide structure, i.e., a field distribution that does not change its shape when propagating along the waveguide. The propagation constant β is a function of the frequency ω; the corresponding relation $\beta = \beta(\omega)$ is referred to as the dispersion relation of the respective waveguide mode. A positive (negative) propagation constant $\beta(\omega) > 0$ ($\beta(\omega) < 0$) corresponds to a propagation of the phase fronts in positive (negative) z-direction.

For a fixed frequency ω, the set of eigenmodes splits up into a discrete number of guided modes and a continuous spectrum of radiation modes [72]. For most practical geometries, the set of guided modes can further be divided into two subsets of different polarization: Modes with a dominant E_x-component are referred to as (quasi-)TE-polarized modes, whereas modes with a dominant E_y-component are called (quasi-)TM-modes. The guided modes we label with a single discrete index μ, whereas the labelling of radiation modes requires two quantities: A continuous parameter β that corresponds to the propagation constant of the mode, and a discrete parameter ζ that includes various field- and symmetry properties [86]. The most general solution consists of a superposition of all these modes,

$$\widetilde{\underline{\mathbf{E}}}(\mathbf{r}, \omega) = \sum_{\mu} A_{\mu} \mathcal{E}_{\mu}(x, y, \omega)\, e^{-j\beta_{\mu}(\omega)z} + \sum_{\zeta} \int_{\beta} A_{\zeta,\beta} \mathcal{E}_{\zeta,\beta}(x, y, \omega)\, e^{-j\beta z}\, d\beta \qquad (1.35)$$

$$\widetilde{\underline{\mathbf{H}}}(\mathbf{r}, \omega) = \sum_{\mu} A_{\mu} \mathcal{H}_{\mu}(x, y, \omega)\, e^{-j\beta_{\mu}(\omega)z} + \sum_{\zeta} \int_{\beta} A_{\zeta,\beta} \mathcal{H}_{\zeta,\beta}(x, y, \omega)\, e^{-j\beta z}\, d\beta. \qquad (1.36)$$

Here A_{μ} is the amplitude of the μ-th guided mode, and $A_{\zeta,\rho}$ is the amplitude density of the radiation modes.

The electric and magnetic fields of the guided and radiation modes form a biorthogonal set, and dropping the space arguments for the ease of notation, the orthogonality relations read [72]

$$\frac{1}{4} \int_{-\infty}^{+\infty} \int_{-\infty}^{+\infty} \left[(\mathcal{E}_{\mu}(\omega) \times \mathcal{H}_{\mu'}^{*}(\omega)) + (\mathcal{E}_{\mu'}^{*}(\omega) \times \mathcal{H}_{\mu}(\omega)) \right] \cdot \mathbf{e}_z\, dx\, dy = \mathcal{P}_{\mu}(\omega)\delta_{\mu,\mu'} \qquad (1.37)$$

$$\frac{1}{4} \int_{-\infty}^{+\infty} \int_{-\infty}^{+\infty} \left[(\mathcal{E}_{\zeta,\beta}(\omega) \times \mathcal{H}_{\zeta',\beta'}^{*}(\omega)) + (\mathcal{E}_{\zeta',\beta'}^{*}(\omega) \times \mathcal{H}_{\zeta,\beta}(\omega)) \right] \cdot \mathbf{e}_z\, dx\, dy = \mathcal{P}_{\zeta,\beta}(\omega)\, \delta_{\zeta,\zeta'}\, \delta(\beta - \beta')$$

$$\qquad (1.38)$$

$$\frac{1}{4} \int_{-\infty}^{+\infty} \int_{-\infty}^{+\infty} \left[(\mathcal{E}_{\zeta,\beta}(\omega) \times \mathcal{H}_{\mu}^{*}(\omega)) + (\mathcal{E}_{\mu}^{*}(\omega) \times \mathcal{H}_{\zeta,\beta}(\omega)) \right] \cdot \mathbf{e}_z\, dx\, dy = 0. \qquad (1.39)$$

Here, $\mathbf{e}_z$ denotes the unit-vector in z-direction, $\delta_{\mu,\mu'}$ is the Kronecker delta, $\delta(\rho - \rho')$ is the Dirac delta function. $\mathcal{P}_{\nu}(\omega)$ $(\mathcal{P}_{\zeta,\rho}(\omega))$ is the power (power density) associated with the corresponding guided (radiation) mode fields[3],

$$\mathcal{P}_{\mu}(\omega) = \frac{1}{2} \int_{-\infty}^{+\infty} \int_{-\infty}^{+\infty} \mathrm{Re}\left\{ \mathcal{E}_{\mu}(\omega) \times \mathcal{H}_{\mu}^{*}(\omega) \right\} \cdot \mathbf{e}_z\, dx\, dy \qquad (1.40)$$

$$\mathcal{P}_{\zeta,\beta}(\omega)\delta(\beta - \beta') = \frac{1}{2} \int_{-\infty}^{+\infty} \int_{-\infty}^{+\infty} \mathrm{Re}\left\{ \mathcal{E}_{\zeta,\beta}(\omega) \times \mathcal{H}_{\zeta,\beta}^{*}(\omega) \right\} \cdot \mathbf{e}_z\, dx\, dy. \qquad (1.41)$$

---

[3]This power is the "numerical" power that is contained in the mode fields. The physical power carried by a certain mode depends on the mode amplitude, $P_{\nu} = |a_{\mu}|^2 \mathcal{P}_{\nu}$.

The calculation of the mode fields $\mathcal{E}(x, y, \omega)$ and $\mathcal{H}(x, y, \omega)$ is not a trivial task, especially for the radiation modes. A vast amount of literature exists on how to calculate guided modes of waveguides with basically arbitrary cross-sections, see [102] for a good overview. Easy-to-apply modesolvers are commercially available, integrated into powerful electromagnetic simulator packages. These modesolvers are mostly based on finite-difference [90, 6] or finite-element [91] schemes.

For the radiation mode fields, however, things are more complicated. These mode fields are not confined to a certain part of the waveguide cross section, but they extend to infinity, which makes their numerical calculation difficult. For simple geometries, e.g. for a 2D slab waveguide [72] or for optical fibers [74], they can be calculated analytically. Only recently, Poulton *et al.* have developed a semi-analytical method based on an expansion in terms of cylindrical harmonics, which allows to calculate the radiation modes for rectangular waveguide geometries [86].

1.2.2 Signal Propagation in Ideal Linear Waveguides

For an ideal waveguide, no coupling between the various guided and radiation modes occurs, and in many practical cases, only a single guided mode of the waveguide is excited. The mode expansion can then be limited to this mode. For the ease of notation, we will drop the mode subscript μ in this section, keeping in mind that the procedure can be applied to any guided mode of an ideal waveguide. We further assume that the signal consists of a carrier wave with frequency $\omega_c \neq 0$ modulated by a slowly-varying envelope. We are interested in how the envelope changes during propagation in positive z-direction. A slowly-varying envelope approximation (SVEA) ansatz is used,

$$\underline{\mathbf{E}}(\mathbf{r}, t) = A(z, t) \frac{\mathcal{E}(x, y, \omega_c)}{\sqrt{P_c(\omega_c)}} \, e^{j(\omega_c t - \beta(\omega_c)z)} , \tag{1.42}$$

$$\underline{\widetilde{\mathbf{E}}}(\mathbf{r}, \omega) = \widetilde{A}(z, \omega - \omega_c) \frac{\mathcal{E}(x, y, \omega_c)}{\sqrt{P_c(\omega_c)}} \, e^{-j \beta(\omega_c)z} , \tag{1.43}$$

Analogous expressions are used for the $\mathbf{H}$-fields,

$$\underline{\mathbf{H}}(\mathbf{r}, t) = A(z, t) \frac{\mathcal{H}(x, y, \omega_c)}{\sqrt{P_c(\omega_c)}} \, e^{j(\omega_c t - \beta(\omega_c)z)} , \tag{1.44}$$

$$\underline{\widetilde{\mathbf{H}}}(\mathbf{r}, \omega) = \widetilde{A}(z, \omega - \omega_c) \frac{\mathcal{H}(x, y, \omega_c)}{\sqrt{P_c(\omega_c)}} \, e^{-j \beta(\omega_c)z} . \tag{1.45}$$

$\mathcal{P}_c(\omega_c)$ is given by Eq. (1.40) and is used for power normalization of the mode field[4]. The physical power carried by the signal is the square of the complex pulse envelope's magnitude,

$$P(z,t) = |A(z,t)|^2 \,. \tag{1.46}$$

Using this ansatz, we inherently assume that the complex transverse mode fields do not change considerably within the bandwidth of the signal and can therefore be approximated by the mode field at the carrier frequency,

$$\mathcal{E}(x,y,\omega) \approx \mathcal{E}(x,y,\omega_c) \quad \text{and} \quad \mathcal{H}(x,y,\omega) \approx \mathcal{H}(x,y,\omega_c) \quad \text{for} \quad \widetilde{A}(z,\omega - \omega_c) \neq 0.$$

This is a reasonable assumption for narrowband signals, i.e., for signals obtained by electric modulation of an optical carrier. The evolution of the pulse envelope can be investigated by introducing Eqs. (1.43) and (1.45) into (1.32) and (1.31). As shown in Appendix C, this leads to

$$\frac{\partial \widetilde{A}(z,\omega - \omega_c)}{\partial z} + \mathrm{j}\left(\beta\left(\omega\right) - \beta(\omega_c)\right) \widetilde{A}(z,\omega - \omega_c) = 0. \tag{1.47}$$

We expand the dispersion relation of the waveguide into a Taylor series around ω_c,

$$\beta(\omega) = \beta_c + \beta_c^{(1)} \cdot (\omega - \omega_c) + \frac{1}{2}\beta_c^{(2)} \cdot (\omega - \omega_c)^2 + \frac{1}{6}\beta_c^{(3)} \cdot (\omega - \omega_c)^3 + \dots \tag{1.48}$$

where $\beta_c^{(n)} = \left.\frac{\mathrm{d}^n \beta}{\mathrm{d}\omega^n}\right|_{\omega=\omega_c}$. Equation (1.47) can then be transformed to the time domain,

$$\frac{\partial A(z,t)}{\partial z} + \beta_c^{(1)} \frac{\partial A(z,t)}{\partial t} - \mathrm{j}\frac{1}{2}\beta_c^{(2)} \frac{\partial^2 A(z,t)}{\partial t^2} - \frac{1}{6}\beta_c^{(3)} \frac{\partial^3 A(z,t)}{\partial t^3} + \dots = 0. \tag{1.49}$$

If the Taylor expansion is terminated after the first-order term, the differential equation can be solved directly. Identifying $\beta_c^{(1)} z$ as the group delay,

$$\tau_g = \beta_c^{(1)} z, \tag{1.50}$$

and $1/\beta_c^{(1)}$ as the group velocity,

$$v_g = \left(\left.\frac{\partial \beta}{\partial \omega}\right|_{\omega=\omega_c} \right)^{-1}, \tag{1.51}$$

we obtain

$$A\left(z,t\right) = A\left(0, t - \frac{z}{v_g}\right). \tag{1.52}$$

Since we have terminated the Taylor expansion (1.48) after the linear term, we find only a translation of the pulse envelope with the group velocity that is associated with

---

[4]The power normalization is good practice, because the numerically calculated mode fields are arbitrary to the extent that they can be multiplied with any complex number. Such a multiplication, however, must not have any consequence on physical power that is associated with a certain mode amplitude. Therefore, the mode fields must be properly normalized when being used in any expansion. Rather than incorporating the power normalization into the definition of the transverse mode fields, we have used an explicit form because this allows us to leave the electric and the magnetic mode field with the usual units V/m and A/m, while at the same time associating the total physical power with $|A(z,t)|^2$.

the carrier frequency. If we take into account higher order terms, we can describe the deformation of the envelope due to chromatic dispersion. The coefficient $\beta_{\mathrm{c}}^{(2)}$ is referred as the group-velocity dispersion (GVD)-parameter, and $\beta_{\mathrm{c}}^{(3)}$ is known as the third-order dispersion (TOD)-parameter. In fiber optics, the dispersion parameter D_2 is commonly used instead of $\beta_{\mathrm{c}}^{(2)}$,

$$D_2 = \left.\frac{\mathrm{d}\beta^{(1)}}{\mathrm{d}\lambda}\right|_{\omega=\omega_{\mathrm{c}}} = -\frac{2\pi c}{\lambda_{\mathrm{c}}^2}\beta_{\mathrm{c}}^{(2)}, \tag{1.53}$$

where $\lambda_{\mathrm{c}} = 2\pi c/\omega_{\mathrm{c}}$ is the wavelength of the carrier.

If the deformation of the pulse envelope is not of interest, $\beta_{\zeta,\mathrm{c}}^{(\nu)} = 0 \ \forall\, \nu > 1$ can be assumed. Furthermore, if a retarded time frame is introduced,

$$t' = t - \beta_{\mathrm{c}}^{(1)} z, \tag{1.54}$$

the first-order partial derivative with respect to z can also be eliminated, and the right-hand side of Eq. (1.49) reduces to

$$\frac{\partial A'(z,t')}{\partial z} = \left.\left[\frac{\partial A(z,t)}{\partial z} + \beta_{\mathrm{c}}^{(1)}\frac{\partial A(z,t)}{\partial t}\right]\right|_{t=t'+\beta_{\mathrm{c}}^{(1)} z}. \tag{1.55}$$

In the following we will generally skip the primes on the time-coordinate, provided that there is no risk of confusion.

1.2.3 Influence of Small Index Profile Changes

If the refractive index profile $n(\mathbf{r})$ of a waveguide is changed by a small amount $\Delta n(\mathbf{r}) \ll n(\mathbf{r})$, the associated change of the propagation constant β_{c} can be calculated by using a perturbation approach. A small refractive index change $\Delta n(\mathbf{r})$ corresponds to a change $\Delta\epsilon(\mathbf{r}) = \epsilon_0\,\Delta\epsilon_r(\mathbf{r})$ in electric permeability, where

$$\Delta\epsilon_r(\mathbf{r}) = \left(n(\mathbf{r}) + \Delta n(\mathbf{r})\right)^2 - n^2(\mathbf{r}) \approx 2\,n(\mathbf{r})\,\Delta n(\mathbf{r})\,. \tag{1.56}$$

Second- and higher-order dispersion is neglected, and a retarded time frame according to Eqs. (1.54) and (1.55) is introduced. The evolution of the pulse envelope is then obtained from Eqs. (C.23) and (C.25),

$$\frac{\partial A_\mu(z,t)}{\partial z} = \mathrm{j}\,\Delta\beta_\mu(z,t)\,A_\mu(z,t), \tag{1.57}$$

where

$$\Delta\beta_\mu(z,t) = \frac{\omega_{\mathrm{c}}}{4P(\omega_{\mathrm{c}})}\iint [\Delta\epsilon_r(\mathbf{r},t)\,\mathcal{E}_\mu(x,y,\omega_{\mathrm{c}})]\cdot\mathcal{E}_\mu^*(x,y,\omega_{\mathrm{c}})\,\mathrm{d}x\,\mathrm{d}y. \tag{1.58}$$

The dot $\cdot$ denotes a scalar product. In many cases of practical interest, the refractive index change is confined to and constant within a certain "active" domain D_{act} (refractive index n_{act}) of the cross section,

$$\Delta n(\mathbf{r},t) = \begin{cases} \Delta n_{\mathrm{act}}(z,t) & \text{inside } D_{\mathrm{act}}\,, \\ 0 & \text{elsewhere.} \end{cases} \tag{1.59}$$

The change $\Delta\beta_\mu(z,t)$ of the propagation constant can then be represented by

$$\Delta\beta_\mu(z,t) = k_0\Gamma_\mu\,\Delta n_{\mathrm{act}}(z,t),\tag{1.60}$$

where the field confinement factor Γ_μ is defined by

$$\Gamma_\mu = \frac{n_{\mathrm{act}}\displaystyle\iint_{D_{\mathrm{act}}}|\mathcal{E}_\mu(x,y,\omega_{\mathrm{c}})|^2\,\mathrm{d}x\,\mathrm{d}y}{Z_0\displaystyle\iint_{D_{\mathrm{tot}}}\mathrm{Re}\left\{\mathcal{E}_\mu(x,y,\omega_{\mathrm{c}})\times\mathcal{H}_\mu^*(x,y,\omega_{\mathrm{c}})\right\}\cdot e_z\,\mathrm{d}x\,\mathrm{d}y}.\tag{1.61}$$

For weakly guiding waveguides, the field confinement factor Γ_μ relates the power guided in the active zone to the total guided power.

1.3   Linear Coupling of Modes in Rough Waveguides

As opposed to the ideal waveguide structure assumed so far, real waveguides exhibit surface roughness, which couples the guided modes to the radiation modes and therefore leads to scattering loss. The surface roughness can again be understood as a perturbation $\Delta\epsilon(\mathbf{r})$ of the ideal waveguide's scalar dielectric profile $\epsilon(\mathbf{r})$. To describe scattering loss mathematically, we use a mode expansion that comprises both guided and radiation modes,

$$\underline{\mathbf{E}}(\mathbf{r},t) = \sum_\zeta \sum\!\!\!\!\!\!\int_{\beta_{\mathrm{c}}} \mathrm{d}\beta_{\mathrm{c}}\, A_{\zeta,\beta_{\mathrm{c}}}(z,t,\omega_{\mathrm{c}})\frac{\mathcal{E}_{\zeta,\beta_{\mathrm{c}}}(x,y,\omega_{\mathrm{c}})}{\sqrt{\mathcal{P}_{\zeta,\beta_{\mathrm{c}}}}}\,\mathrm{e}^{-\mathrm{j}(\omega_{\mathrm{c}}t-\beta_{\mathrm{c}}z)},\tag{1.62}$$

$$\underline{\mathbf{H}}(\mathbf{r},t) = \sum_\zeta \sum\!\!\!\!\!\!\int_{\beta_{\mathrm{c}}} \mathrm{d}\beta_{\mathrm{c}}\, A_{\zeta,\beta_{\mathrm{c}}}(z,t,\omega_{\mathrm{c}})\frac{\mathcal{H}_{\zeta,\beta_{\mathrm{c}}}(x,y,\omega_{\mathrm{c}})}{\sqrt{\mathcal{P}_{\zeta,\beta_{\mathrm{c}}}}}\,\mathrm{e}^{-\mathrm{j}(\omega_{\mathrm{c}}t-\beta_{\mathrm{c}}z)}.\tag{1.63}$$

The symbol $\sum\!\!\!\!\int$ is to be understood as a sum over the discrete propagation constants $\beta_{\mathrm{c},\nu}$ of the guided modes plus an integral over the continuum of the propagation constants β_{c} that belong to the various radiation modes at the carrier frequency ω_{c}. The sum $\sum_\zeta$ covers all possible mode symmetries that belong to a certain values of β_{c}. Second- and higher-order dispersion is neglected, and a retarded time frame according to Eqs. (1.54) and (1.55) is introduced. According to Eq. (C.22), the evolution of the pulse envelope is then given by

$$\frac{\partial A_{\zeta',\beta_{\mathrm{c}}'}(z,t)}{\partial z} = -\mathrm{j}\sum_\zeta \sum\!\!\!\!\!\!\int_{\beta_{\mathrm{c}}} \mathrm{d}\beta_{\mathrm{c}}K_{(\zeta',\beta_{\mathrm{c}}')(\zeta,\beta_{\mathrm{c}})}(z)\,A_{\zeta,\beta_{\mathrm{c}}}(z,t)\,\mathrm{e}^{-\mathrm{j}(\beta_{\mathrm{c}}-\beta_{\mathrm{c}}')z},\tag{1.64}$$

The mode coupling coefficient $K_{(\zeta',\beta_{\mathrm{c}}')(\zeta,\beta_{\mathrm{c}})}(z)$ is obtained from an overlap integral of the refractive index perturbation $\Delta\epsilon(\mathbf{r})$ in the respective cross section, see Eq. (C.23),

$$K_{(\zeta',\beta_{\mathrm{c}}')(\zeta,\beta_{\mathrm{c}})}(z) = \frac{\omega_{\mathrm{c}}}{4\sqrt{\mathcal{P}_{\zeta',\beta_{\mathrm{c}}'}\mathcal{P}_{\zeta,\beta_{\mathrm{c}}}}}\iint\left(\Delta\epsilon(\mathbf{r})\mathcal{E}_{\zeta,\beta_{\mathrm{c}}}(x,y,\omega_{\mathrm{c}})\cdot\mathcal{E}_{\zeta',\beta'}^*(x,y,\omega_{\mathrm{c}})\right)\mathrm{d}x\,\mathrm{d}y.\tag{1.65}$$

For rough waveguides, $K_{(\zeta',\beta'_c)(\zeta,\beta_c)}(z)$ is a weakly stationary stochastic process. The scattering loss depends on the statistics of this process, in particular on the spatial power spectrum of the sidewall perturbations. This will be discussed in more detail in Chapter 3.

1.4 Electro-Optic Interaction

For linear electro-optic materials, a microwave field can change the refractive index at optical frequencies. The microwave field has a carrier frequency of $\omega_0 = 0$ and the corresponding electric field is denoted as[5]

$$\mathbf{E}_{\mathrm{mw}}(\mathbf{r}, t) = A(z, t, \omega_0)\frac{\mathcal{E}(x, y, \omega_0)}{\sqrt{\mathcal{P}(\omega_0)}}. \tag{1.66}$$

We will now consider the nonlinear interaction between the microwave field and an optical field $\underline{\mathbf{E}}_{\mathrm{opt}}(\mathbf{r}, t)$ at carrier frequency ω_{c},

$$\underline{\mathbf{E}}_{\mathrm{opt}}(\mathbf{r}, t) = A(z, t, \omega_{\mathrm{c}})\frac{\mathcal{E}(x, y, \omega_{\mathrm{c}})}{\sqrt{\mathcal{P}(\omega_{\mathrm{c}})}}\, \mathrm{e}^{-\mathrm{j}\beta(\omega_{\mathrm{c}})z}. \tag{1.67}$$

Due to the linear electro-optic effect (Pockels effect), the microwave field changes the refractive index seen by the various field components of the optical mode. The corresponding time- and space-dependent change $\Delta\beta(z, t)$ of the optical mode's propagation constant $\beta(\omega_{\mathrm{c}})$ can be deduced from Eqs. (C.33), (C.34), and (C.36),

$$\Delta\beta(z, t) = \frac{\omega_{\mathrm{c}}}{2\mathcal{P}(\omega_{\mathrm{c}})}\iint \left[\underline{\widetilde{\chi}}^{(2)}(\omega_{\mathrm{c}} : 0, \omega_{\mathrm{c}}) : \mathbf{E}_{\mathrm{mw}}(\mathbf{r}, t)\mathcal{E}(x, y, \omega_{\mathrm{c}})\right] \cdot \mathcal{E}^*(x, y, \omega_{\mathrm{c}})\,\mathrm{d}x\,\mathrm{d}y. \tag{1.68}$$

In an equivalent formulation, the microwave field $\underline{\mathbf{E}}_{\mathrm{mw}}(\mathbf{r}, t)$ induces a time- and space-dependent change $\Delta\underline{\epsilon}_r(\mathbf{r}, t)$ of the permeability tensor $\underline{\epsilon}_r(\mathbf{r}, t)$,

$$\underline{\epsilon}_r(\mathbf{r}, t) = \underline{\epsilon}_{r,0}(\mathbf{r}) + \Delta\underline{\epsilon}_r(\mathbf{r}, t). \tag{1.69}$$

Since electro-optic materials are always anisotropic, $\underline{\epsilon}_r(\mathbf{r}, t)$ is a 3×3 matrix that connects the optical field $\underline{\mathbf{E}}_{\mathrm{opt}}(\mathbf{r}, t)$ and the optical displacement $\underline{\mathbf{D}}_{\mathrm{opt}}(\mathbf{r}, t) = \epsilon_0\underline{\epsilon}_r(\mathbf{r}, t)\underline{\mathbf{E}}_{\mathrm{opt}}(\mathbf{r}, t)$. Using Eqs. (1.6), (1.10) and (B.26), and taking into account the appropriate degeneracy factors, Eq. (B.29), we find the relation between $\Delta\underline{\epsilon}_r(\mathbf{r}, t)$, $\underline{\widetilde{\chi}}^{(2)}(\omega_{\mathrm{c}} : 0, \omega_{\mathrm{c}})$, and $\mathbf{E}_{\mathrm{mw}}(\mathbf{r}, t)$,

$$\Delta\underline{\epsilon}_r(\mathbf{r}, t)\mathcal{E}(x, y, \omega_{\mathrm{c}}) = 2\underline{\widetilde{\chi}}^{(2)}(\omega_{\mathrm{c}} : 0, \omega_{\mathrm{c}}) : \mathbf{E}_{\mathrm{mw}}(\mathbf{r}, t)\mathcal{E}(x, y, \omega_{\mathrm{c}}). \tag{1.70}$$

Equation 1.68 can thus be rewritten as

$$\Delta\beta(z, t) = \frac{\omega_{\mathrm{c}}}{4\mathcal{P}(\omega_{\mathrm{c}})}\iint [\Delta\underline{\epsilon}_r(\mathbf{r}, t)\mathcal{E}(x, y, \omega_{\mathrm{c}})] \cdot \mathcal{E}^*(x, y, \omega_{\mathrm{c}})\,\mathrm{d}x\,\mathrm{d}y. \tag{1.71}$$

---

[5]For a signal with zero carrier frequency $\omega_o = 0$, the envelope $A(z, t, \omega_o)$ and the mode field $\mathcal{E}(x, y, \omega_0)$ are real.

This formula is consistent with introducing $\Delta\underline{\epsilon}_r(\mathbf{r}, t)$ directly into Eq. (C.23) and (C.25).

Material data of electro-optic materials are often supplied in terms of electro-optic coefficients r_{mn}. To understand the definition of the electro-optic coefficients, we have to introduce the impermeability tensor η, which links the electric field $\underline{\mathbf{E}}_{\mathrm{opt}}$ to the electric displacement $\underline{\mathbf{D}}_{\mathrm{opt}}$,

$$\underline{\mathbf{E}}_{\mathrm{opt}}(\mathbf{r}, t) = \frac{1}{\epsilon_0}\underline{\eta}(\mathbf{r}, t)\underline{\mathbf{D}}_{\mathrm{opt}}(\mathbf{r}, t). \tag{1.72}$$

The impermeability tensor η can be represented as a 3×3 matrix, which is the inverse of the permeability matrix,

$$\underline{\eta} = \underline{\epsilon}_r^{-1} \tag{1.73}$$

In the presence of a microwave field, the impermeability tensor of a linear electro-optic material is changed by $\Delta\eta(\mathbf{r}, t)$,

$$\underline{\eta}(\mathbf{r}, t) = \underline{\eta}_0(\mathbf{r}) + \Delta\underline{\eta}(\mathbf{r}, t). \tag{1.74}$$

For small changes $\Delta\eta(\mathbf{r}, t)$ of the impermeability tensor, the corresponding change of the permeability tensor is given by

$$\Delta\underline{\epsilon}_r(\mathbf{r}, t) = -\underline{\epsilon}_{r,0}(\mathbf{r})\Delta\underline{\eta}(\mathbf{r}, t)\underline{\epsilon}_{r,0}(\mathbf{r}). \tag{1.75}$$

The permeability tensor $\underline{\epsilon}_r$ is assumed to be real and symmetric for electro-optic materials. The same holds true for its inverse η. Therefore, $\Delta\eta$ contains six independent coefficients: $\Delta\eta_{11}$, $\Delta\eta_{22}$, $\Delta\eta_{33}$, $\Delta\eta_{23} = \Delta\eta_{23}$, $\Delta\eta_{13} = \Delta\eta_{31}$, $\Delta\eta_{12} = \Delta\eta_{21}$. They are connected to the microwave field $\mathbf{E}_{\mathrm{mw}}(\mathbf{r}, t)$ by the electro-optic coefficients r_{mn}, which are generally specified with respect to a material (e. g., crystallographic) coordinate system u, v, w,

$$\begin{pmatrix} \Delta\eta_{11} \\ \Delta\eta_{22} \\ \Delta\eta_{33} \\ \Delta\eta_{23} \\ \Delta\eta_{13} \\ \Delta\eta_{12} \end{pmatrix} = \begin{pmatrix} r_{11} & r_{12} & r_{13} \\ r_{21} & r_{22} & r_{23} \\ r_{31} & r_{32} & r_{33} \\ r_{41} & r_{42} & r_{43} \\ r_{51} & r_{52} & r_{53} \\ r_{61} & r_{62} & r_{63} \end{pmatrix} \begin{pmatrix} E_{\mathrm{mw},u} \\ E_{\mathrm{mw},v} \\ E_{\mathrm{mw},w} \end{pmatrix}. \tag{1.76}$$

The space and time dependence has been omitted for the sake of readability. Crystallographic symmetries can considerably reduce the number of nonzero coefficients [14].

Eqs. (1.68), (1.75) and (1.76) thus permit to calculate the change of the optical propagation constant β that is associated with a certain microwave field $\mathbf{E}_{\mathrm{mw}}(\mathbf{r}, t)$.

1.5 Third-Order Nonlinear Interaction

Third-order optical nonlinearities allow for interaction of signals with nearly identical carrier frequencies and can hence be exploited for ultra-fast all-optical signal processing and wavelength conversion.

In this section, we consider the most prominent cases of third-order nonlinear interaction in waveguides: Self-phase modulation (SPM), cross-phase modulation (XPM), and four-wave mixing (FWM). The formal mathematical derivation of the equations is given in Appendix C, leading to a general relation for third-order nonlinear interaction of signals with different carrier frequencies, propagating in different waveguide modes, see Eq. (C.38). For the further analysis, impulse deformation due to second- and higher-order dispersion is neglected, $\beta^{(1)} \neq 0$, $\beta^{(2)} = \beta^{(3)} = \beta^{(4)} = \cdots = 0$, and retarded time frame according to Eq. (1.55) is used. All signals are assumed to propagate within the same mode, and the mode indices μ_1, μ_2, μ_3 and μ' are omitted for the ease of notation.

1.5.1 Self-Phase Modulation

We first consider third-order nonlinear interaction of a signal with itself, leading to self-phase modulation (SPM). Assuming a signal envelope $A(z, t, \omega_c)$ according to Eqs. (1.42) and (1.44), we have to deal with only one carrier frequency $\omega_1 = \omega_c$ and its negative counterpart $\omega_{-1} = -\omega_c$. Being interested in the change of the amplitude $A(z, t, \omega_\Sigma = \omega_c)$, we have to sum over three triples of frequency indices on the right-hand-side of Eq. (C.38), $\mathbb{S}_{\omega_\nu} = \{(-1, 1, 1), (1, -1, 1), (1, 1, -1)\}$ according to Eq. (C.37). The phase mismatch due to second-order waveguide dispersion vanishes, see Eq. (C.40). We obtain the nonlinear Schrödinger (NLS) equation,

$$\frac{\partial A(z, t, \omega_c)}{\partial z} + \beta^{(1)}(\omega_c) \frac{\partial A(z, t, \omega_c)}{\partial t} - \mathrm{j} \frac{1}{2} \beta^{(2)}(\omega_c) \frac{\partial^2 A(z, t, \omega_c)}{\partial t^2} = -\mathrm{j}\,\gamma \left| A(z, t, \omega_c) \right|^2 A(z, t, \omega_c)$$

$$(1.77)$$

where the nonlinear parameter γ is obtained from Eq. (C.39)

$$\gamma = \frac{3\omega_c \epsilon_0}{16\,\mathcal{P}^2} \iint \left[\underline{\widetilde{\chi}}^{(3)}\left(\omega_c : \omega_c, \omega_c, -\omega_c\right) \vdots \mathcal{E}\left(\omega_c\right) \mathcal{E}\left(\omega_c\right) \mathcal{E}^*\left(\omega_c\right) \right] \cdot \mathcal{E}^*(\omega_c)\, \mathrm{d}x\, \mathrm{d}y. \qquad (1.78)$$

The spatial arguments (x, y) have been omitted. The quantity $\underline{\widetilde{\chi}}^{(3)}$ is the frequency-domain representation of the nonlinear susceptibility tensor.

1.5.2 Effective Area of Nonlinear Interaction

For many cases of practical interest, only the core or the cover material have a $\chi^{(3)}$-nonlinearity, which is usually isotropic. The third-order nonlinear susceptibility tensor $\underline{\chi}^{(3)}$ can then assumed to be zero outside a nonlinear interaction domain D_{inter} (refractive index n_{inter}), and it is nonzero and constant inside D_{inter}. Further, $\underline{\widetilde{\chi}}^{(3)}$ may be approximated by a scalar $\widetilde{\chi}^{(3)}$, so that $\underline{\widetilde{\chi}}^{(3)} \vdots \mathcal{E}\mathcal{E}\mathcal{E}^* = \widetilde{\chi}^{(3)} \left| \mathcal{E} \right|^2 \mathcal{E}$ holds. In the following, $D_{\text{tot}} \supseteq D_{\text{inter}}$ denotes the total cross section of the waveguide.

To evaluate the effects of the waveguide geometry only, the strength of the nonlinear interaction of the guided modes can then be compared to a plane wave in bulk nonlinear material with the same nonlinear susceptibility $\chi^{(3)}$ and the same refractive index as D_{inter}. This leads to the concept of an effective nonlinear interaction area A_{eff}: In a waveguide with a nonlinear interaction region D_{inter} the cross-sectional power P is transported. Relating P to the effective area A_{eff} leads to an effective intensity $I = P/A_{\text{eff}}$. To achieve the same strength of nonlinear interaction, this intensity I should be attributed to a plane wave which propagates in a homogeneous medium with the same optical properties as seen in D_{inter}. The nonlinear waveguide parameter γ can then be written as

$$\gamma = \frac{3\omega_c \epsilon_0 Z_0^2}{4 A_{\text{eff}}\, n_{\text{inter}}^2}\, \widetilde{\chi}^{(3)},\tag{1.79}$$

where the effective area A_{eff} is given by

$$A_{\text{eff}} = \frac{Z_0^2}{n_{\text{inter}}^2}\, \frac{\left| \iint_{D_{\text{tot}}} \operatorname{Re}\left\{ \mathcal{E}(x,y,\omega_c) \times \mathcal{H}^*(x,y,\omega_c) \right\} \cdot \mathbf{e}_z \; \mathrm{d}x\; \mathrm{d}y \right|^2}{\iint_{D_{\text{inter}}} \left| \mathcal{E}(x,y,\omega_c) \right|^4 \mathrm{d}x\; \mathrm{d}y}.\tag{1.80}$$

For complex values of $\widetilde{\chi}^{(3)}$ the nonlinear parameter γ will be also complex, and parametric $\chi^{(3)}$-processes (e.g. SPM, XPM, FWM) will be impaired by nonparametric processes (e.g. TPA).

For low index-contrast material systems, the approximation $n_{\text{core}} \approx n_{\text{cover}} \approx n_{\text{inter}}$ holds, and the longitudinal field components become negligible. The transverse components of the mode fields $\mathcal{E}(x,y)$ and $\mathcal{H}(x,y)$ may then be approximated by a scalar function $F(x,y)$, $\mathcal{E}(x,y) \approx F(x,y)\,\mathbf{e}_x$, $\mathcal{H}(x,y) \approx \frac{n_{\text{inter}}}{Z_0} F(x,y)\,\mathbf{e}_y$. If we further assume a homogeneous nonlinearity, then $D_{\text{inter}} = D_{\text{tot}}$, and Eq. (1.80) can be simplified to

$$A_{\text{eff}} \approx \frac{\left(\iint_{D_{\text{tot}}} |F(x,y)|^2 \, \mathrm{d}x\; \mathrm{d}y \right)^2}{\iint_{D_{\text{tot}}} |F(x,y)|^4 \, \mathrm{d}x\; \mathrm{d}y}.\tag{1.81}$$

This relation is identical with the usual definition of an effective area A_{eff} [3, Eq. (2.3.29)].

1.5.3 Cross-Phase Modulation and Four-Wave Mixing

Nonlinear interaction between two optical signals of different carrier frequencies leads to cross-phase modulation (XPM) and degenerate four wave mixing (FWM). In the following, we consider the interaction of a strong pump wave at carrier frequency ω_p and a weaker signal wave at frequency ω_s. Degenerate FWM leads to the formation of a converted wave at frequency $\omega_c = 2\omega_p - \omega_s$, which, in a "small signal" approximation, is assumed to be weak compared to the signal wave,

$$|A(z,t,\omega_p)|^2 \ll |A(z,t,\omega_s)|^2 \ll |A(z,t,\omega_c)|^2.\tag{1.82}$$

Since the three frequencies ω_p, ω_s, and ω_c are very similar, the corresponding mode fields can be assumed to be identical,

$$\mathcal{E}(x,y) = \mathcal{E}(x,y,\omega_p) \approx \mathcal{E}(x,y,\omega_s) \approx \mathcal{E}(x,y,\omega_c),\tag{1.83}$$

$$\mathcal{H}(x,y) = \mathcal{H}(x,y,\omega_p) \approx \mathcal{H}(x,y,\omega_s) \approx \mathcal{H}(x,y,\omega_c).\tag{1.84}$$

The three waves are subject to linear propagation loss due to scattering at the rough waveguide surface and/or due to material absorption. This is accounted for by a linear attenuation coefficient α_1.

Application of Eq. (C.38), yields three coupled differential equations. Keeping only the dominant terms and dropping the arguments (z, t), we obtain

$$\frac{\partial A(\omega_p)}{\partial z} = -\frac{\alpha_1}{2} A(\omega_p) - \underbrace{\mathrm{j}\,\gamma\,|A(\omega_p)|^2\,A(\omega_p)}_{\text{SPM}}, \tag{1.85}$$

$$\frac{\partial A(\omega_s)}{\partial z} = -\frac{\alpha_1}{2} A(\omega_s) - \underbrace{\mathrm{j}\,2\gamma\,|A(\omega_p)|^2\,A(\omega_p)}_{\text{XPM}}, \tag{1.86}$$

$$\frac{\partial A(\omega_c)}{\partial z} = -\frac{\alpha_1}{2} A(\omega_c) - \underbrace{\mathrm{j}\,2\gamma\,|A(\omega_p)|^2\,A(\omega_c)}_{\text{XPM}} - \underbrace{\mathrm{j}\,\gamma A^2(\omega_p) A^*(\omega_s)\,\mathrm{e}^{-\mathrm{j}\,\Delta\beta z}}_{\text{FWM}}. \tag{1.87}$$

The expressions labelled with SPM, XPM and FWM are responsible for self phase modulation, cross phase modulation and degenerate four wave mixing, respectively. The quantity $\Delta\beta$ accounts for mismatch due to second-order waveguide dispersion[6],

$$\Delta\beta = 2\beta(\omega_p) - \beta(\omega_s) - \beta(\omega_c) \approx -\beta^{(2)}(\omega_p)\,(\omega_s - \omega_p)^2. \tag{1.88}$$

Let us now investigate the four-wave mixing (FWM) conversion efficiency of a waveguide of physical length L,

$$\eta_{\text{FWM}} = \frac{|A(L, t, \omega_c)|^2}{|A(0, t, \omega_s)|^2}. \tag{1.89}$$

The self-phase modulation (SPM) and XPM effects in Eqs. (1.85), (1.85), and (1.85) can be taken into account by modifying the propagation constant. Neglecting signal depletion, the small-signal conversion efficiency can be obtained by integrating Eq. (1.85) over the waveguide length L,

$$\eta_{\text{FWM, ss}} = \mathrm{e}^{-\alpha_1 L} \cdot (\gamma P_{p0} L_{\text{eff}})^2. \tag{1.90}$$

$P_p = |A(0, t, \omega_p)|^2$ denotes the input pump power and the effective length $L_{\text{eff}} < L$ of the waveguide accounts for phase mismatch and linear loss,

$$L_{\text{eff}} = \frac{\sqrt{1 + \mathrm{e}^{-2\alpha_1 L} - 2\,\mathrm{e}^{-\alpha_1 L}\cos(\Delta\beta L)}}{\sqrt{\alpha_1^2 + \Delta\beta^2}}. \tag{1.91}$$

---

[6]Even though the impulse deformation due to second-order waveguide dispersion was neglected, it has to be taken into account here: For dispersion-induced impulse deformation, the bandwidth of the impulse envelope is relevant, whereas phase mismatch in degenerate FWM is dictated by the much larger frequency difference $|\omega_p - \omega_s|$.

Chapter 2

Experimental Setups and Measurement Techniques

2.1   Introduction

The experimental characterization of integrated optical devices is challenging in different ways: First, the width of high index-contrast waveguides is typically a fraction of a micrometer. Coupling light from an optical fiber to these waveguides thus requires precise and stable mechanical positioning of the fiber tips with respect to the facet of the device under test. Second, the modes of high index-contrast waveguides have typically sub-μm diameter, and thus may exhibit considerable mode mismatch even to lensed fibers. Coupling losses can thus not be neglected and must be carefully characterized to gain insight into the on-chip power levels. Third, high index-contrast integrated optical devices are polarization sensitive. Therefore light must be coupled to and from these devices in a defined polarization state, which should be stable over the entire measurement. Fourth, the behaviour of integrated optical devices is mostly very sensitive to temperature variations. The device mount must therefore be temperature-controlled, especially if active devices are investigated. Lastly, the response times of integrated optical devices are generally too small to be resolved even by state-of-the-art electronics. For linear devices, the analysis can be done in the frequency domain using high-precision laser sources that can be tuned over a wide wavelength range. For nonlinear devices, however, dynamics must be investigated in the time domain, using pump-probe techniques with ultrashort laser pulses.

In the following sections, we present different experimental setups that allow for the static and the dynamic characterization of linear and nonlinear devices. These techniques are crucial for the measurements described in the subsequent chapters.

2.2   Device Mount and Fiber Positioning

Figure 2.1 shows a photograph of a device mount for integrated optical devices. The device under test (1) is carried by a device holder (2), which is kept at a constant temperature using an electronically controlled Peltier element (3) mounted directly underneath. From

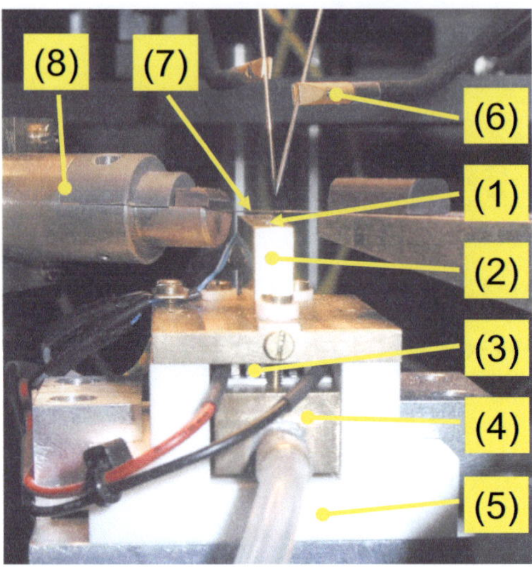

Fig. 2.1. Photograph of the designed device mount. (1) DUT; (2) Device holder; (3) Peltier element; (4) Heat sink; (5) Base; (6) Contact needles; (7) Lensed fiber tip; (8) Rotary mount for polarization-maintaining fiber tips

the lower side of the Peltier element, the heat is directly drained into a water-cooled heat sink (4). The heat sink is integrated into the base (5) of device mount in a way that allows for expansion or shrinkage without influencing the position of the device under test (DUT). For active devices, pump currents are supplied to the DUT by using probe needles (6) attached to micropositioners. Light is coupled to and from the DUT using lensed fiber tips (7) that are positioned by differential micrometer screws and by feedback-controlled piezoelectric actuators (not shown in Fig. 2.1). To ensure that the a well-defined linear polarization state is launched to the DUT, polarization maintaining lensed fibers are used at least at the input side. In the following, TE (TM) refers to a linearly-polarized excitation for which the dominant component of the electric field is parallel (perpendicular) to the plane of the optical table. By using special mounts (8) that allow to rotate the fiber tips around their longitudinal axes, the tips are aligned such that the slow (fast) axes is parallel (perpendicular) to the optical table. Coupling light into the slow (fast) axis of the lensed fibers thus results in TE (TM) excitation of the DUT.

In the following sections, the equipment for mounting and cooling the DUT, for positioning the fibers, and for supplying the current is taken for granted. For the sake of clarity, this equipment is omitted in Figs. 2.2, 2.3 and 2.4. The same holds true for any computer-controlled processing of measurement data.

2.3   Spectral Characterization of Linear Devices

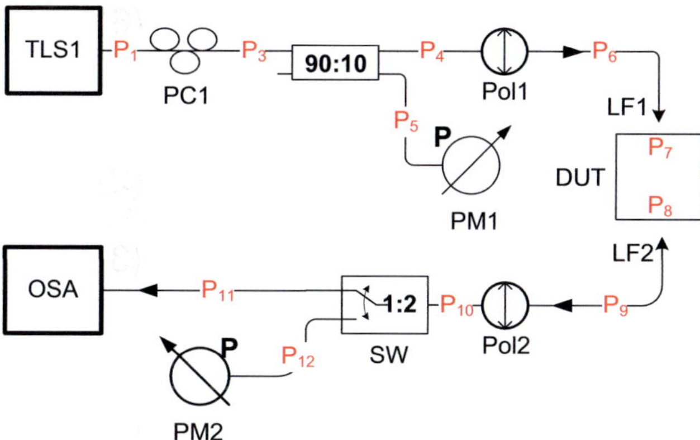

Fig. 2.2. Setup for spectral characterization of linear devices. TLS1 = tunable laser source; PC1 = polarization controller; Pol1, Pol2 = polarizers; PM1, PM2 = power meters; LF1, LF2 = polarization-maintaining lensed fibers; SW = switch; OSA = optical spectrum analyzer. The symbols P_1, P_2, ... P_{12} denote optical power levels at different points of the setup.

Figure 2.2 shows the setup used for spectral characterization of linear devices. A tunable laser source TLS1 (tunable wavelength range 1480 nm to 1580 nm, resolution 1 pm) provides narrowband (linewidth < 1 MHz) light of optical power $P_1 \leq 7$ dBm in a single-mode fiber. A certain fraction P_5 of the power is tapped later for control of the power level. Using a polarization controller PC1, the polarization can be adjusted to the transmission axis of a polarizer Pol1. The lensed fibers LF1 and LF2 are polarization maintaining. If the transmission axis of Pol1 is aligned parallel to the slow (fast) axis of LF1, TE (TM) light is launched into the DUT. Similarly, the TE (TM) polarization can be investigated at the output of the DUT by aligning the transmission axis of Pol2 parallel to the slow (fast) axis of LF2. Using a 1×2 optical switch SW, the light collected from the DUT can either be coupled to a power meter PM2 or to an optical spectrum analyzer OSA (resolution 15 pm, level range -90 dBm to $+20$ dBm). As a special feature of the measurement hardware, the tunable laser source and the optical spectrum analyzer allow for synchronized wavelength sweep. This permits narrow-band (low-noise) spectral characterization of active devices even in the presence of a strong amplified spontaneous emission (ASE) background.

For analyzing the absolute power levels in the measurement system, the insertion loss of the optical components has to be quantified. For this purpose we use ratios $a_{\nu,\mu}$ of the various power levels P_μ, P_ν depicted in Fig. 2.2,

$$a_{\nu,\mu} = 10 \log_{10}\left(\frac{P_\mu}{P_\nu}\right) \text{ dB.} \tag{2.1}$$

At a wavelength of 1550 nm, the transmission of the polarizers was found to be $a_{6,4} \approx -2.85$ dB and $a_{10,9} \approx -4.65$ dB. The transmission of the switch is typically $a_{11,10} \approx -0.8$ dB, and for a device based on InGaAsP/InP pedestal waveguides, a fiber-to-fiber transmission of $a_{9,6} = -13.7$ dB was measured. Neglecting the on-chip propagation loss $a_{8,7}$, an upper bound for the coupling efficiency can thus be estimated to be $a_{7,6} \approx a_{9,8} \approx a_{9,6}/2 = -6.8$ dB.

The setup for spectral characterization of linear devices was used to the propagation loss of rough waveguides (Chapter 3), the insertion loss of waveguide bends (Chapter 4), and the linear transmission spectra of microring resonators (Chapter 8).

2.4   Static Characterization of Nonlinear Devices

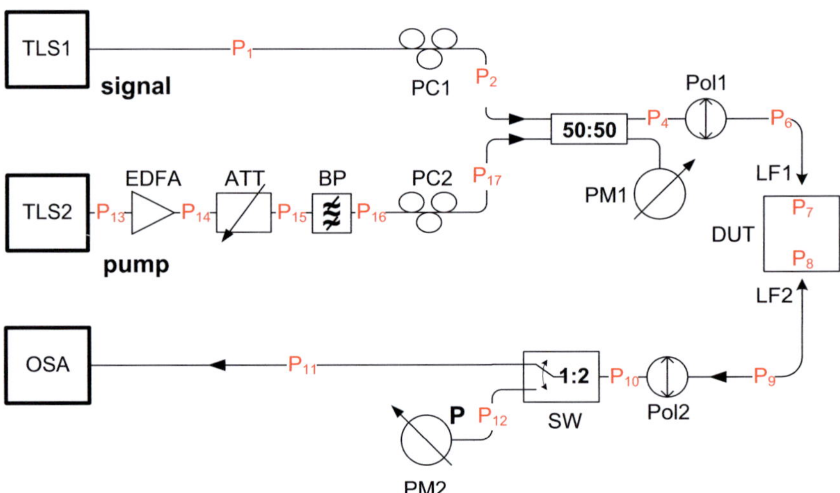

Fig. 2.3. Setup for static characterization of nonlinear devices. A high-power pump wave and the signal wave are coupled to the device under test (DUT). TLS1, TLS2 = tunable laser sources; EDFA = erbium-doped fiber amplifier; ATT = optical attenuator; BP = optical band-pass filter; PC1, PC2 = polarization controllers; Pol1, Pol2 = polarizers; PM1, PM2 = power meters; LF1, LF2 = lensed fibers; SW = switch; OSA = optical spectrum analyzer. The symbols P_1, P_2, ... , P_{17} denote optical power levels at different points of the setup.

Figure 2.3 depicts a setup used for characterizing nonlinear properties of devices operated at high continuous wave (cw) power levels. This is done by coupling a high-power pump wave at frequency ω_p to the device together with the signal at frequency ω_s. For third-order nonlinear devices, degenerate four-wave mixing leads to the formation of converted wave at frequency $\omega_c = 2\omega_p - \omega_s$.

In contrast to the setup used for spectral characterization of linear devices, Fig. 2.2, the setup in Fig. 2.3 comprises a second narrowband tunable laser source TLS2 (tunable wavelength range 1460 nm to 1580 nm, resolution 0.1 pm, output power $\leq$ 6 dBm, linewidth 100 kHz), amplified by a two-stage high-power erbium-doped fiber amplifier (EDFA, optical bandwidth 1535 nm to 1565 nm, saturated output power +33 dBm, noise figure < 7 dB). The EDFA is always operated at high power levels where it exhibits low noise figure, and the power level coupled into the DUT can be adjusted using an optical attenuator ATT. An optical band-pass filter BP is used to reduce the ASE background originating from the EDFA. The polarization of the pump wave can be adjusted to the transmitting axis of Pol1 using the polarization controller PC2. At the output side, the setup is identical to the one discussed in the last section.

The setup was used for the experimental investigation of degenerate four-wave mixing (FWM) in passive InGaAsP/InP microring resonators (Chapter 9).

2.5 Dynamic Pump-Probe Characterization of Nonlinear Devices

2.5.1 Measurement Principle

Pump-probe techniques are the workhorse for time-domain characterization of ultrafast nonlinear optical devices. A change in the transmission characteristics of a waveguide induced by nonlinear interaction with a high-power pump impulse is sampled by a weak probe impulse. By varying the delay between the pump and the probe impulse, the dynamics of the nonlinearity can be investigated.

A technique that allows for using pump and probe impulses of identical polarizations and center wavelengths has been described in [46]. In this scheme, the probe impulse is marked by a frequency shift, and is detected by measuring amplitude and phase of a beat signal between the pump signal and a reference signal. Due to its similarity to heterodyne detection of optical signals, this method is also referred as heterodyne pump-probe technique.

In the following sections, we describe a heterodyne pump-probe setup that was constructed within the framework of this thesis. The setup was used for the dynamic characterization of a nonlinear InAs/GaAs quantum dot (QD) amplifier at 1300 nm (Chapter 8).

2.5.2 Experimental Setup

Figure (2.4) shows a sketch of the heterodyne pump-probe setup that was constructed and used in the framework of this thesis. The femtosecond pulse train is obtained from a cascade of solid-state lasers: High-power laser diodes ($P_0 \approx 40$ W) are used to pump a frequency-doubled neodymium-doped yttrium vanadate (Nd:YVO$_4$) laser (NdYV). This cw laser emits approximately $P_1 = 9.4$ W of green light ($\lambda_{\mathrm{NdYV}} = 532$ nm), which is used to pump a mode-locked femtosecond laser (MLL) with titanium-doped sapphire as gain medium. The MLL generates a train of impulses with typically $\Delta t_{\mathrm{MLL,FWHM}} < 130$ fs full

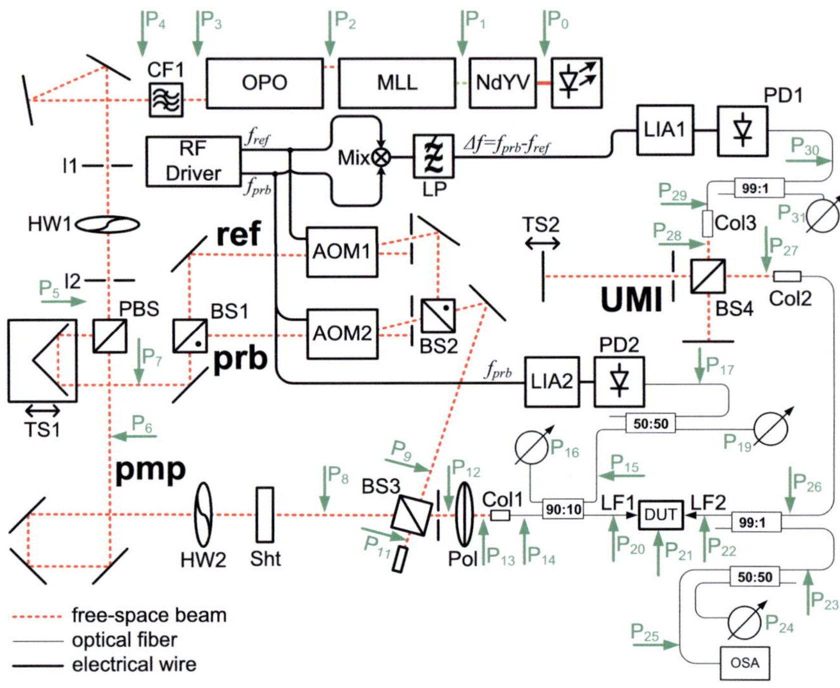

Fig. 2.4. Heterodyne pump-probe setup. Free-space optical beams are indicated by dashed lines, optical fibers are represented by thin continuous lines, and thick continuous lines refer to electrical wires. The pulse train is split up into a pump path (pmp), a probe path (prb) and a reference path (ref). Using acousto-optic modulators, the probe and the reference impulses are shifted in frequency by f_{prb} and f_{ref}, respectively. The probe and the reference impulses are overlayed by passing through a Michelson inteferometer of unbalanced arm lengths (UMI), and the beat signal is measured using lock-in detection at the difference frequency $\Delta f = f_{prb} - f_{ref}$. At the same time, the relative position of the probe and the pump impulse is obtained by measuring their cross correlation. NdYV = neodymium yttrium vanadate (Nd:YVO$_4$) pump laser; MLL = titanium-sapphire modelocked femtosecond laser; OPO = optical parametric amplifier; CF1 = color filter; I1, I2 = iris diaphragms; HW1, HW2 = half-wave plates; PBS = polarizing beam splitter; TS1, TS2 = translation stages; BS1, BS2, BS3, BS4 = non-polarizing beam splitters, AOM1, AOM2 = acousto-optic modulators; Sht = beam shutter; Pol = polarizer; Col1, Col2, Col3 = fiber collimators; LF1, LF2 = lensed fibers; DUT = device under test; OSA = optical spectrum analyzer; UMI = unbalanced Michelson interferometer; PD1, PD2 = photodetectors; LIA1, LIA2 = lock-in amplifiers; Mix = electrical mixer; LP = electrical low-pass filter. The symbols P_1, P_2, ..., P_{31} denote optical power levels at different points of the setup.

width at half the maximum (FWHM). The average output power is $P_2 = 1.8\,\mathrm{W}$, and the ultra-stable repetition frequency is $f_{\mathrm{rep}} = 80.4785\,\mathrm{MHz}$. The center wavelength λ_{MLL} can be tuned from $720\,\mathrm{nm}$ to $850\,\mathrm{nm}$. This train of impulses is used to pump a femtosecond optical parametric oscillator (OPO) which is synchronized to the repetition frequency of the MLL by automatic control of the cavity length. By exploiting a nonlinear three-wave interaction in a lithium triborate (LBO) crystal, the pump wave at λ_{MLL} is transformed into a signal wave at λ_{OPO} and an idler wave at λ_{idl}, where $1/\lambda_{\mathrm{MLL}} = 1/\lambda_{\mathrm{OPO}} + 1/\lambda_{\mathrm{idl}}$. For $\lambda_{\mathrm{OPO}} = 1550\,\mathrm{nm}$ ($\lambda_{\mathrm{OPO}} = 1300\,\mathrm{nm}$), a pump wavelength of $\lambda_{\mathrm{MLL}} = 810\,\mathrm{nm}$ ($\lambda_{\mathrm{MLL}} = 775\,\mathrm{nm}$) is required, and the temperature of the LBO crystal has to be tuned to the respective optimum. Using nonlinear autocorrelation techniques, a FWHM impulse width of $\Delta t_{\mathrm{OPO,FWHM}} = 120\,\mathrm{fs}$ ($\Delta t_{\mathrm{OPO,FWHM}} = 150\,\mathrm{fs}$) is typically measured for the output signal of the OPO at $\lambda_{\mathrm{OPO}} = 1550\,\mathrm{nm}$ ($\lambda_{\mathrm{OPO}} = 1300\,\mathrm{nm}$). The average output power typically amounts to $P_3 = 240\,\mathrm{mW}$ ($P_3 = 270\,\mathrm{mW}$).

The pulse train from the OPO is first filtered by a color filter CF1 to remove spurious idler and pump light. Two irises I1 and I2 are used to realign the position of the free-space beam after adjusting the pulsed laser system. The pulse train is then split up into two parts. The power ratio P_6/P_7 can be tuned using a half-wave plate HW1 and a polarizing beam splitter PBS.

The power P_6 transmitted by the PBS is used as a pump beam and is labelled with pmp. It passes through a half-wave plate HW2, which, together with the polarizer Pol allows to adjust the pump power launched into the fiber collimator Col1. For reference measurements, the pump beam can be blanked by means of a computer-controlled shutter Sht.

The reflected power P_7 first passes through a adjustable delay line that consists of a corner cube mounted onto a translation stage TS1 and is then split up further by a nonpolarizing beamsplitter BS1 into probe (prb) path and a reference path (ref). The light passing through the probe (reference) path is up-shifted in frequency by f_{prb} (f_{ref}) using an acousto-optic modulator AOM1 (AOM2). The probe and the reference beams are recombined in a nonpolarizing beamsplitter BS2, whereby the reference impulse now exhibits a fixed delay τ_2 with respect to the probe impulse. The probe/reference ensemble is then recombined with the pump beam by a non-polarizing beamsplitter BS3, whereby the delays are chosen such that the reference impulse is always ahead of the same period's probe impulse[1]. The train of reference, pump and probe impulse and the respective delays are depicted in Fig. 2.5 (a). The delay τ_1 between the pump and the probe impulses can be adjusted by moving the translation stage TS1, whereas the delay τ_2 between the probe and the reference impulses is fixed.

The combined train of reference, pump and probe impulses then passes through a polarizer POL and is coupled into a polarization-maintaining fiber using a collimator Col1. A certain fraction P_{15} of the power in the fiber is tapped for monitoring purpose. The remaining power P_{20} is launched onto the input facet of the DUT. At the output of the DUT, again a small fraction P_{23} of the optical power is again tapped for wavelength and power monitoring. The remainder passes through the collimator Col2 and enters a free-beam unbalanced Michelson interferometer (UMI). At the output of the UMI, two

---

[1]To avoid nonlinear effects induced by the leading reference impulse, the power of the reference signal must be chosen significantly smaller than the power of the pump signal.

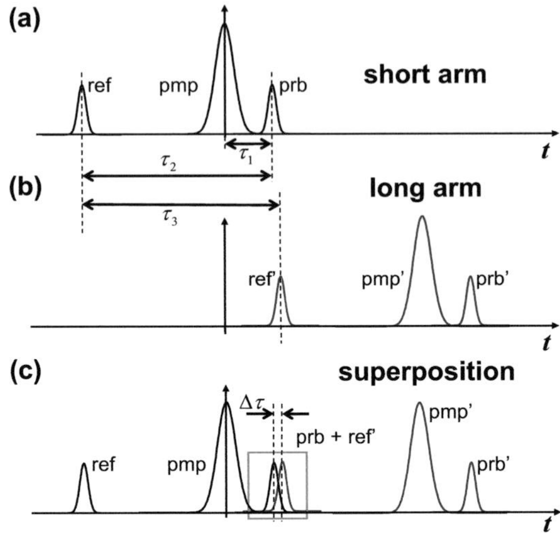

Fig. 2.5. (a) Impulse sequence obtained from the short arm of the unbalanced Michelson interferometer (UMI). The delay τ_1 between the probe (prb) and the pump (pmp) impulse can be adjusted by moving the translation stage TLS1. The reference impulse is delayed by a fixed amount $\tau_2 \approx 800$ ps with respect to the probe impulse. The carrier frequency of the reference impulse (ref) is up-shifted by f_{ref} and the carrier frequency of the probe impulse (ref) is up-shifted by f_{prb} with respect to the carrier frequency f_c of the pump impulse (pmp). (b) Corresponding impulse sequence obtained from the long arm of the UMI. The relative delay due to unbalanced arm length is denoted as τ_3, and the delayed reference, pump and probe impulses are labelled ref', pmp' and prb', respectively. (c) Superposition of impulse sequences after the UMI. The temporal overlap of probe impulse prb (carrier frequency $f_c + f_{\mathrm{prb}}$) and reference impulse ref' (carrier frequency $f_c + f_{\mathrm{ref}}$) generates a beat signal at $\Delta f = f_{\mathrm{prb}} - f_{\mathrm{ref}}$.

replica of the impulse ensembles are overlaid with a relative time delay τ_3, see Figs. 2.5 (a) and (b). Using a translation stage TS2, τ_3 can be adjusted such that the trailing probe impulse of the short arm is overlaid with the leading reference impulse of the long arm , see Fig. 2.5 (c). At the output side of the UMI, the signal is collected by a fiber collimator Col3. Again, a small fraction P_{31} of the optical power is tapped and measured, whereas the remaining power P_{30} is impinged on a low-noise photodetector PD1. A lock-in amplifier synchronized to the beat frequency $\Delta f = f_{\mathrm{prb}} - f_{\mathrm{ref}}$ of the overlapping probe and reference impulses is used to extract complex amplitude of the beat signal. The magnitude of the beat signal is directly related to the amplitude of the probe impulses, and the phase of the beat signal corresponds to the phase difference between the probe and the reference impulses. The mathematical relationship shall be derived in the next section.

For correct interpretation of the measurement results, it is important to know the pump power and the position of TS1 which corresponds to zero pump-probe delay. Therefore the power P_{15} (tapped before the DUT) is split up into two branches P_{17} and P_{19}, the latter of which is used to measure the pump power. The signal corresponding to P_{17} is impinged on a photodetector PD2, and the frequency component at f_{prb} is extracted from the photocurrent by a lock-in amplifier LIA2. As shall be shown in the next section, this allows to measure the cross-correlation of the pump and the probe impulses and gives valuable information on the width and the relative position of these impulses.

2.5.3 Mathematical Formulation

To describe the measurement principle in a mathematical way, impulse envelopes at different points along the beam path have to be considered. In the following, the z-coordinate is measured along the beam path, and $z = z_\mu$ corresponds to the point at which the power P_μ is defined, see Fig. 2.4. We use a retarded time frame according to Eq. (1.54), and we define that $t = 0$ corresponds to the moment in which the pump impulse (of the short arm of the UMI) arrives at $z = z_\mu$, see Fig. 2.5 (a). The different path lengths that the reference (ref), pump (pmp) and probe (prb) impulses have to travel through the system are expressed by the time delays τ_1, τ_2, and τ_3. The delay τ_1 of the probe with respect to the pump impulse can be varied between $-130\,\mathrm{ps}$ (probe impulse ahead of pump impulse) to $670\,\mathrm{ps}$ by moving the translation stage TLS1. The reference impulse is delayed by a fixed amount $\tau_2 \approx 800\,\mathrm{ps}$ with respect to the probe impulse.

In the following, $\tau_{rep} = 1/f_{rep}$ is the period of the impulse trains, $\omega_c = 2\pi c/\lambda_{OPO}$ is the optical carrier frequency, and $\omega_{ref} = 2\pi f_{ref}$ ($\omega_{prb} = 2\pi f_{prb}$) denotes the angular frequency by which AOM1 (AOM2) shifts up the reference (probe) signal. $A_{ref}(z, t)$, $A_{pmp}(z, t)$ and $A_{prb}(z, t)$ are the slowly-varying envelopes of a single reference, pump and probe impulse, respectively.

Heterodyne Detection of the Probe Signal

We will first consider the overlaid impulse sequence after the UMI and calculate the signal generated by PD1. After passing through the short arm of the UMI, the reference, pump and probe pulse trains at $z = z_{30}$ can be written as

$$a_{ref}(z_{30}, t) = \sum_{\nu=-\infty}^{\infty} A_{ref}\left(z_{30}, t - \tau_1 + \tau_2 - \nu\tau_{rep}\right) e^{j\left((\omega_c + \omega_{ref})(t - \tau_1 + \tau_2) + \varphi_{ref}\right)}, \tag{2.2}$$

$$a_{pmp}(z_{30}, t) = \sum_{\nu=-\infty}^{\infty} A_{pmp}\left(z_{30}, t - \nu\tau_{rep}\right) e^{j(\omega_c t + \varphi_{pmp})}, \tag{2.3}$$

$$a_{prb}(z_{30}, t) = \sum_{\nu=-\infty}^{\infty} A_{prb}\left(z_{30}, t - \tau_1 - \nu\tau_{rep}\right) e^{j\left((\omega_c + \omega_{prb})(t - \tau_1) + \varphi_{prb}\right)}, \tag{2.4}$$

see Fig. 2.5 (a). The corresponding pulse trains for the long arms are obtained by replacing t with $t - \tau_3$ in Eqs. 2.2, 2.3 and 2.4, see Fig. 2.5 (b). Omitting the argument z_{30}, the total optical signal at $z = z_{30}$ thus reads

$$a_{tot}(t) = a_{ref}(t) + a_{pmp}(t) + a_{prb}(t) + a_{ref}(t - \tau_3) + a_{pmp}(t - \tau_3) + a_{prb}(t - \tau_3). \tag{2.5}$$

The associated optical power $P(t)$ averaged over some optical cycles is given by

$$P_{\text{tot}}(t) = |a_{\text{tot}}(t)|^2 .\tag{2.6}$$

The electrical bandwidth of PD1 is far smaller than the repetition frequency f_{rep} of the pulse trains, i.e., PD1 averages over many pulse cycles. PD1 comprises a transimpedance amplifier, which transforms the photocurrent $i(t) = SP_{\text{tot}}(t)$ into a voltage $u(t) = Gi(t)$, where S is the responsivity and G denotes the transimpedance gain (unit: V / A). The electrical signal measured at the output of PD1 is thus given by

$$u(t) = \frac{GS}{\tau_{\text{int}}} \int\limits_{t-\tau_{\text{int}}/2}^{t+\tau_{\text{int}}/2} \left|a_{\text{tot}}(\bar{t})\right|^2 \mathrm{d}\bar{t}.\tag{2.7}$$

The integral extends over a period of length τ_{int}, where τ_{int} is the integration time associated with the finite electrical bandwidth of PD1. The integral typically extends over a couple cycles τ_{rep} of the pulse trains.

The signal $u(t)$ contains a spectral component $u_{\Delta\omega}(t)$, centered at the difference frequency $\Delta\omega = \omega_{\text{prb}} - \omega_{\text{ref}}$. This spectral component corresponds to the beat of the frequency-shifted probe and reference impulses[2],

$$u_{\Delta\omega}(t) = GS \left[\frac{1}{\tau_{\text{rep}}} \int\limits_{-\infty}^{+\infty} A_{\text{prb}}(z_{30}, \bar{t})\, A^*_{\text{ref}}(z_{30}, \bar{t} - \Delta\tau)\, \mathrm{d}\bar{t} \right] e^{j\,\Delta\omega\, t}\, e^{j(\omega_c+\omega_{\text{ref}})\,\Delta\tau}\, e^{j\,\Delta\varphi}\tag{2.8}$$

The quantity $\Delta\varphi = \varphi_{\text{prb}} - \varphi_{\text{ref}} - \Delta\omega\tau_2$ denotes a constant phase difference between the probe and the reference impulse, and $\Delta\tau = \tau_3 - \tau_2$ is the residual delay mismatch between the probe and the reference impulse. The lock-in amplifier LIA1 extracts the complex amplitude U of $u_{\Delta\omega}(t)$, thereby averaging over many pulse cycles τ_{rep},

$$U = \left[\frac{GS}{\tau_{\text{rep}}} \int\limits_{-\infty}^{+\infty} A_{\text{prb}}(z_{30}, \bar{t})\, A^*_{\text{ref}}(z_{30}, \bar{t} - \Delta\tau)\, \mathrm{d}\bar{t} \right] e^{j(\omega_c+\omega_{\text{ref}})\,\Delta\tau}\, e^{j\,\Delta\varphi} .\tag{2.9}$$

Reference Measurements

The strong pump impulse induces a nonlinear change of the DUT's properties. The dynamics of the pump-induced transmission are then sampled with the probe and the reference impulses. Both of them are weak enough to ensure that they do not induced any further nonlinearities. If the pump impulses do not change, the DUT can thus be assumed to transmit the probe and the reference impulses in a linear, but time-dependent way.

---

[2] $A_{\text{prb}}(z_{30}, t)$ and $A_{\text{ref}}(z_{30}, t)$ refer to the envelopes of a single probe and reference impulse. These impulses are much shorter than the pulse repetition period τ_{rep}. The integrand of Eq. (2.8) hence decays quickly for $|t| \to \infty$, and the boundaries of the integration range do not matter. For simplicity, we have set them to $\pm\infty$ rather than to $\pm\tau_{\text{rep}}/2$.

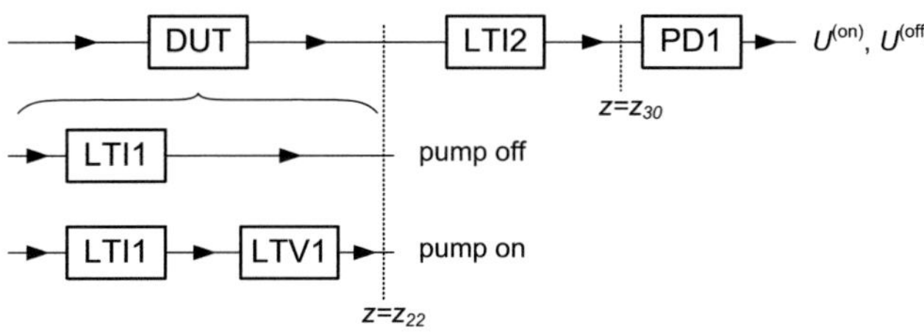

Fig. 2.6. Block diagram of the measurement system; DUT = device under test, PD1 = photodetector, see Fig. (2.4); LTI1, LTI2 = linear time-invariant systems; LTV = linear time-variant system. LTI2 comprises all components in between the DUT and PD1, e.g., fibers, couplers, collimators, and the UMI, see Fig. (2.4). The DUT itself is split up into a linear time invariant system LTI1 and a linear time-variant system LTV1.

To characterize the nonlinear dynamics of the DUT, the pump-probe delay τ_1 is scanned, and the pump-induced changes of the DUT's transmission are measured. For each value of τ_1, the following procedure is completed: First, the pump beam is blocked by means of the shutter Sht, and the complex amplitude $U^{(\mathrm{off})}$ of the difference frequency component $u_{\Delta\omega}(t)$ is measured by LIA1. Then the pump beam is cleared by opening the shutter, and the amplitude $U^{(\mathrm{on})}$ is measured. Then the ratio $R = U^{(\mathrm{on})}/U^{(\mathrm{off})}$ is calculated and stored.

In the following, $A_{\mathrm{prb}}^{(\mathrm{on})}(z_{30}, t)$ and $A_{\mathrm{ref}}^{(\mathrm{on})}(z_{30}, t)$ denote the slowly-varying envelopes of the probe and the reference impulse at $z = z_{30}$ with the pump switched on, and $A_{\mathrm{prb}}^{(\mathrm{off})}(z_{30}, t)$ and $A_{\mathrm{ref}}^{(\mathrm{off})}(z_{30}, t)$ are envelopes for the pump beam switched off. According to Eq. (2.9), the corresponding complex voltage amplitudes $U^{(\mathrm{on})}$ and $U^{(\mathrm{off})}$ can be written as

$$U^{(\mathrm{on})} = \left[\frac{GS}{\tau_{\mathrm{rep}}} \int_{-\infty}^{+\infty} A_{\mathrm{prb}}^{(\mathrm{on})}(z_{30}, \bar{t}) \left[A_{\mathrm{ref}}^{(\mathrm{on})}(z_{30}, \bar{t} - \Delta\tau) \right]^{*} d\bar{t} \right] e^{\mathrm{j}(\omega_c + \omega_{\mathrm{ref}})\Delta\tau} e^{\mathrm{j}\Delta\varphi}, \qquad (2.10)$$

$$U^{(\mathrm{off})} = \left[\frac{GS}{\tau_{\mathrm{rep}}} \int_{-\infty}^{+\infty} A_{\mathrm{prb}}^{(\mathrm{off})}(z_{30}, \bar{t}) \left[A_{\mathrm{ref}}^{(\mathrm{off})}(z_{30}, \bar{t} - \Delta\tau) \right]^{*} d\bar{t} \right] e^{\mathrm{j}(\omega_c + \omega_{\mathrm{ref}})\Delta\tau} e^{\mathrm{j}\Delta\varphi}. \qquad (2.11)$$

For the pump beam turned on, the magnitude and the phase of the probe impulse envelope $A_{\mathrm{prb}}^{(\mathrm{on})}(t)$ clearly depends on the pump-probe-delay τ_1, and so do $U^{(\mathrm{on})}$ and the ratio $R = U^{(\mathrm{on})}/U^{(\mathrm{off})}$. This dependence shall now be investigated in more detail.

System Model

Figure 2.6 shows a block diagram of the relevant parts of the measurement system: DUT and PD1 refer to the device under test and to the photodetector PD1, see Fig. (2.4). All components in between the DUT and PD1, e.g., fibers, couplers, collimators, and the UMI, see Fig. (2.4), are represented by a linear time-invariant system LTI2. Given the fact that the components of LTI2 have constant insertion loss over the frequency range of interest, the corresponding transfer $\tilde{h}_{\mathrm{LTI2}}(\omega)$ function may be approximated by $\tilde{h}_{\mathrm{LTI2}}(\omega) \approx \sqrt{P_{30}/P_{22}}\exp(\mathrm{j}\,\varphi_{\mathrm{LTI2}}(\omega))$, where the power levels P_{22} and P_{30} are defined in Fig. (2.4) and where dispersion is represented by the frequency-dependent phase $\varphi_{\mathrm{LTI2}}(\omega)$. Using this approximation, it can be shown that LTI2 does not have any influence on the ratio $R = U^{(\mathrm{on})}/U^{(\mathrm{off})}$. The impulse envelopes $A_{\mathrm{prb}}^{(\mathrm{on})}(z_{30}, t)$ and $A_{\mathrm{ref}}^{(\mathrm{on})}(z_{30}, t)$ at $z = z_{30}$ in Eqs. (2.10) and (2.17), can therefore be replaced by their counterparts $A_{\mathrm{prb}}^{(\mathrm{on})}(z_{22}, t)$ and $A_{\mathrm{ref}}^{(\mathrm{on})}(z_{22}, t)$ at $z = z_{22}$, i.e. at the output of the DUT. LTI2 may therefore be disregarded for the further analysis. We will drop the z-dependence for the sake of clearer notation, and from now on, all envelope functions $A_{\mathrm{prb}}^{(\mathrm{on})}(t)$, $A_{\mathrm{ref}}^{(\mathrm{on})}(t)$ refer to $z = z_{22}$.

To facilitate the analysis, the DUT itself is split up into another linear time-invariant (LTI) system LTI1 and a subsequent linear time-variant (LTV) system LTV1. LTI1 contains all transmission properties of the DUT that are not time-dependent, in particular the insertion loss and the waveguide dispersion of the DUT. The LTV part comprises the pump-induced time-dependence. If the pump beam is turned off, the output of the LTI system $A_{\mathrm{prb}}^{(\mathrm{off})}(t)$ and $A_{\mathrm{ref}}^{(\mathrm{off})}(t)$ is directly measured. For the pump beam turned on, the signals $A_{\mathrm{prb}}^{(\mathrm{off})}(t)$ and $A_{\mathrm{ref}}^{(\mathrm{off})}(t)$ pass through the subsequent LTV system, and the outputs $A_{\mathrm{prb}}^{(\mathrm{on})}(t)$ and $A_{\mathrm{ref}}^{(\mathrm{on})}(t)$ are measured. Using the Green's function representation of a linear time-variant system according to Eq. (A.17), the pump-induced changes of the probe and the reference beam can be described by integration with a time-dependent kernel. At $z = z_{22}$, the impulse envelopes with and without pump are thus related by

$$A_{\mathrm{prb}}^{(\mathrm{on})}(t - \tau_1) = \int\limits_{-\infty}^{+\infty} A_{\mathrm{prb}}^{(\mathrm{off})}(\tau - \tau_1) g(t - \tau, \tau)\, \mathrm{d}\tau, \tag{2.12}$$

$$A_{\mathrm{ref}}^{(\mathrm{on})}(t - \tau_1 + \tau_2) = \int\limits_{-\infty}^{+\infty} A_{\mathrm{ref}}^{(\mathrm{off})}(\tau - \tau_1 + \tau_2) g(t - \tau, \tau)\, \mathrm{d}\tau. \tag{2.13}$$

The kernel $g(t - \tau, \tau)$ in Eqs. (2.12) and (2.13) refers only to the LTV part of the DUT and depends in a nonlinear way on the shape and the energy of the pump impulse.

Let us now consider the LTV part of the DUT in more detail: The LTI part already accounts for all effects that lead to a delay and a deformation of the impulses within the DUT. i.e., the "linear memory" of the waveguide can be absorbed into the LTI part. The LTV-part can then be assumed to respond instantaneously to an excitation,

$$g(t - \tau, \tau) = T(\tau)\,\mathrm{e}^{\mathrm{j}\,\phi(\tau)}\,\delta(t - \tau). \tag{2.14}$$

Here $T(\tau)$ denotes the time-dependent amplitude transmission and $\phi(\tau)$ is the time-dependent phase shift. Eqs. (2.12) and (2.13) can thus be simplified

$$A_{\mathrm{prb}}^{(\mathrm{on})}(t) = A_{\mathrm{prb}}^{(\mathrm{off})}(t) T(t + \tau_1)\,\mathrm{e}^{\mathrm{j}\,\phi(t+\tau_1)}, \tag{2.15}$$

$$A_{\mathrm{ref}}^{(\mathrm{on})}(t) = A_{\mathrm{ref}}^{(\mathrm{off})}(t) T(t + \tau_1 - \tau_2)\,\mathrm{e}^{\mathrm{j}\,\phi(t+\tau_1-\tau_2)}. \tag{2.16}$$

The corresponding complex voltage amplitudes $U^{(\mathrm{on})}$ for the pump beam turned on is thus given by

$$U^{(\text{on})} = \frac{GS}{\tau_{\text{rep}}} \, \mathrm{e}^{\mathrm{j}(\omega_c + \omega_{\text{ref}})\,\Delta\tau + \Delta\varphi} \tag{2.17}$$

$$\times \int_{-\infty}^{+\infty} A_{\text{prb}}^{(\text{off})}(\bar{t}) T(\bar{t} + \tau_1) \, \mathrm{e}^{\mathrm{j}\,\phi(\bar{t}+\tau_1)} \Big[A_{\text{ref}}^{(\text{off})}(\bar{t} - \Delta\tau) T(\bar{t} - \Delta\tau + \tau_1 - \tau_2) \, \mathrm{e}^{\mathrm{j}\,\phi(\bar{t}-\Delta\tau+\tau_1-\tau_2)} \Big]^* \mathrm{d}\bar{t}$$

Temporal Resolution

For short laser impulses, the probe and reference envelopes $A_{\text{prb}}(\bar{t})$, $A_{\text{ref}}(\bar{t})$ are different from zero only in a small region around $\bar{t} = 0$. For the weak reference impulse, which enters the DUT before the strong pump impulse ($\tau_1 - \tau_2 \ll 0$), the transmission of the DUT may therefore be assumed constant,

$$A_{\text{ref}}^{(\text{off})}(t-\Delta\tau) T(t-\Delta\tau+\tau_1-\tau_2) \, \mathrm{e}^{\mathrm{j}\,\phi(t-\Delta\tau+\tau_1-\tau_2)} \approx A_{\text{ref}}^{(\text{off})}(t-\Delta\tau) T(\tau_1-\tau_2) \, \mathrm{e}^{\mathrm{j}\,\phi(\tau_1-\tau_2)} . \tag{2.18}$$

The probe impulse, however, experiences the fast response of the DUT. For pulse widths that are comparable to the response times of the DUT, the measured device response is blurred due to nonzero probe impulse width. Defining a window function $w(t)$,

$$w(t) = \frac{A_{\text{prb}}^{(\text{off})}(t) \left[A_{\text{ref}}^{(\text{off})}(t - \Delta\tau) \right]^*}{\displaystyle\int_{-\infty}^{+\infty} A_{\text{prb}}^{(\text{off})}(\bar{t}) \left[A_{\text{ref}}^{(\text{off})}(\bar{t} - \Delta\tau) \right]^* \mathrm{d}\bar{t}}, \tag{2.19}$$

Eq. (2.17) can be simplified,

$$U^{(\text{on})} = \frac{GS}{\tau_{\text{rep}}} \, \mathrm{e}^{\mathrm{j}(\omega_c + \omega_{\text{ref}})\,\Delta\tau + \Delta\varphi} \, T(\tau_1 - \tau_2) \, \mathrm{e}^{-\mathrm{j}\,\phi(\tau_1-\tau_2)} \tag{2.20}$$

$$\times \int_{-\infty}^{+\infty} w(\bar{t}) T(\bar{t} + \tau_1) \, \mathrm{e}^{\mathrm{j}\,\phi(\bar{t}+\tau_1)} \, \mathrm{d}\bar{t} \int_{-\infty}^{+\infty} A_{\text{prb}}^{(\text{off})}(\bar{t}) \left[A_{\text{ref}}^{(\text{off})}(\bar{t} - \Delta\tau) \right]^* \mathrm{d}\bar{t}.$$

The ratio $R = U^{(\text{on})}/U^{(\text{off})}$ reveals the response function of the LTV system, blurred by convolution with $w(-t)$,

$$R(\tau_1) = T(\tau_1 - \tau_2) \, \mathrm{e}^{-\mathrm{j}\,\phi(\tau_1-\tau_2)} \int_{-\infty}^{+\infty} w(\bar{t}) T(\bar{t} + \tau_1) \, \mathrm{e}^{\mathrm{j}\,\phi(t+\tau_1)} \, \mathrm{d}\bar{t} \tag{2.21}$$

If the response times of the DUT are much longer than the width of the window function $w(t)$, blurring can be neglected, and the product of the probe and the reference transmission is obtained directly,

$$R(\tau_1) = T(\tau_1) \, \mathrm{e}^{\mathrm{j}\,\phi(\tau_1)} \, T(\tau_1 - \tau_2) \, \mathrm{e}^{-\mathrm{j}\,\phi(\tau_1-\tau_2)} . \tag{2.22}$$

The detuning $\Delta\tau$ of the UMI may slightly reduce the temporal resolution by broadening $w(t)$, but does not severely impair the measurement result.

Interpretation of Measurement Results

The pump power coupled into the DUT is periodic, $|a_{\mathrm{pmp}}(t)|^2 = |a_{\mathrm{pmp}}(t - \tau_{\mathrm{rep}})|^2$, which makes the Green's function $g(t - \tau, \tau)$ of the DUT periodic in τ,

$$g(t - \tau, \tau) = g(t - (\tau + \tau_{\mathrm{rep}}), \tau + \tau_{\mathrm{rep}}). \tag{2.23}$$

The period $\tau_{\mathrm{rep}} = 1/f_{\mathrm{rep}}$ is of the order of $12\,\mathrm{ns}$. Fast nonlinear processes in the DUT will recover between two subsequent pump impulses, but slow processes, e.g., thermal effects will accumulate over many periods and finally settle to a steady-state background. This has to be taken into account when interpreting the measured data $R(\tau_1)$.

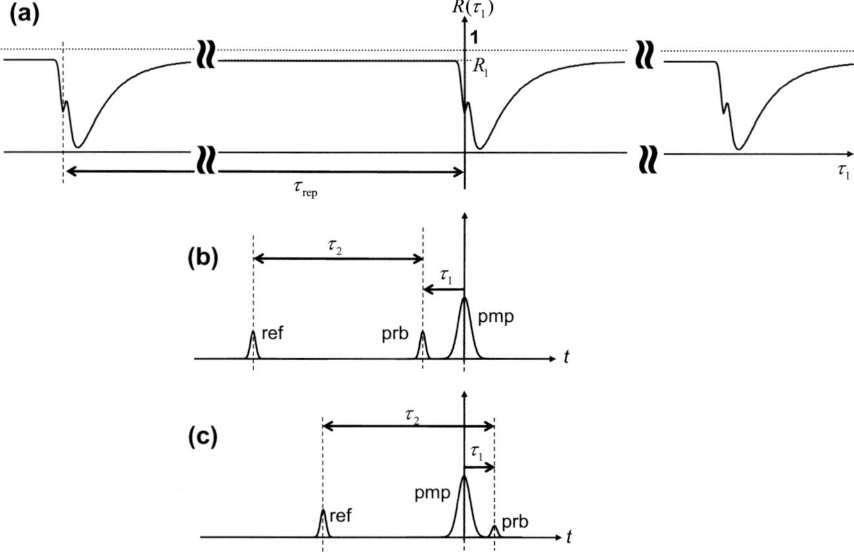

Fig. 2.7. (a) Signal amplitude ratio $R(\tau_1) = U^{(\mathrm{on})}/U^{(\mathrm{off})}$ as a function of pump-probe delay τ_1. $R(\tau_1)$ is periodic in τ_1 with period τ_{rep}. (b) For sufficiently negative pump-probe delay $\tau_1 \ll 0$, both the probe impulse and the reference impulse are ahead of the same period's pump impulse. Both of them are thus influenced by slow processes that have accumulated over the previous pump periods. This leads to a nearly constant value $R_1 = |T_1|^2$. (c) For pump-probe delays that are approximately zero or positive, $\tau_1 \geq 0$, the trailing probe impulse is influenced by fast pump-induced nonlinear processes in the DUT, whereas the reference impulse can still be assumed to experience the steady-state amplitude transmission.

The measured complex ratio $R(\tau_1)$ is also periodic in τ_1 with period τ_{rep}, but only a small fraction of the period can be sampled within $\tau_1 = -130\,\mathrm{ps} \ldots 670\,\mathrm{ps}$. This is depicted in Fig. (2.7) (a).

We will first consider the case $\tau_1 \ll 0$, where "$\tau_1 \ll 0$" indicates that τ_1 must be sufficiently negative to prevent even partial overlap of the pump and the probe impulses.

In this case both the probe impulse and the reference impulse are ahead of the same period's pump impulse, see Fig. (2.7) (b). The probe and the reference impulses are thus only influenced by slow processes that have accumulated over the previous pump periods. This leads to a nearly constant value of the DUT transmission which influences the probe and the reference impulse in equal measure, $T(\tau_1 - \tau_2) \approx T(\tau_1) \approx T_1$ and $\phi(\tau_1 - \tau_2) \approx \phi(\tau_1) = \phi_1$ for $-130\,\mathrm{ps} < \tau_1 \ll 0$. The measured voltage ratio for $-130\,\mathrm{ps} < \tau_1 \ll 0$ is thus

$$R(\tau_1) = R_1 = T_1^2. \tag{2.24}$$

The accumulated steady-state transmission $T_1 = \sqrt{R_1}$ can thus be directly calculated from R_1.

For $\tau_1 \geq 0$, only the probe impulse is influenced by fast pump-induced nonlinear processes in the DUT, whereas the reference impulse can still be assumed to experience the steady-state amplitude transmission T_1 and the corresponding phase shift ϕ_1, see Fig. (2.7) (c). Eq. (2.21) can thus be written as

$$R(\tau_1) = T_1 \int\limits_{-\infty}^{+\infty} w(\bar{t}) T(\bar{t} + \tau_1)\, e^{j(\phi(\bar{t}+\tau_1) - \phi_1)}\, \mathrm{d}\bar{t}. \tag{2.25}$$

The blurred ultrafast response function can thus be obtained by computing the ratio $R(\tau_1)/T_1$. If the blurring is neglected, the response of the DUT is obtained directly,

$$T(\tau_1)\, e^{j\,\Delta\phi(\tau_1)} = \frac{R(\tau_1)}{T_1} \tag{2.26}$$

where $\Delta\phi(\tau_1) = \phi(\tau_1) - \phi_1$ is the phase shift introduced by the ultrafast nonlinear interaction between the pump and the probe impulse.

For DUT dynamics that are "slow" compared to the width of the probe impulse, the heterodyne pump-probe setup allows to measure the response function of the DUT directly, see Eq. (2.26). If the response times of the DUT are on the order of the probe impulse width, the temporal resolution can be improved by deconvolving according to Eq. (2.25).

Measurement of the Pump-Probe Cross-Correlation

The cross-correlation of the pump and the probe impulse contains important information about the relative positions and the temporal overlap of the two impulses. In particular, from the maximum of the cross correlation, we can find the position of the translation stage TS1 for which the probe and the pump impulse coincide, and the width of the cross-correlation can at least give a clue of the impulse widths[3].

The pump-probe cross-correlation is measured as a function of pump-probe delay τ_1 by using a photodetector PD2 and extracting the pump-probe beat at frequency f_prb using

---

[3]Note that the width of the cross-correlation is not directly linked to the widths of the individual impulses. As a matter of fact, the cross-correlation does not show dispersion-induced impulse broadening that is common to both the pump and the probe impulse.

a lock-in amplifier LIA2. The complex amplitude of the pump and the probe pulse trains at PD2 ($z = z_{17}$) can be written as

$$a_{\mathrm{pmp}}(z_{17}, t) = \sum_{\nu=-\infty}^{\infty} A_{\mathrm{pmp}}(z_{17}, t - \nu \tau_{\mathrm{rep}}) \, e^{j(\omega_c t + \varphi_{\mathrm{pmp}})} , \tag{2.27}$$

$$a_{\mathrm{prb}}(z_{17}, t) = \sum_{\nu=-\infty}^{\infty} A_{\mathrm{prb}}(z_{17}, t - \tau_1 - \nu \tau_{\mathrm{rep}}) \, e^{j\big((\omega_c + \omega_{\mathrm{prb}})(t - \tau_1) + \varphi_{\mathrm{prb}}\big)} . \tag{2.28}$$

The optical power, averaged over some periods of the carrier wave, is then given by $P_{\mathrm{PD2}}(t) = |a_{\mathrm{pmp}}(z_{17}, t) + a_{\mathrm{prb}}(z_{17}, t)|^2$. The bandwidth of PD2 must be sufficient to show the beat signal between pump and probe at frequency f_{prb}. Given the responsivity S_{PD2} and the transimpedance gain G_{PD2} of PD2, and taking into account that LIA2 averages over many pulse cycles τ_{rep}, the complex voltage amplitude $U_{\mathrm{PD2},\omega_{\mathrm{prb}}}$ is given by

$$U_{\mathrm{PD2},\omega_{\mathrm{prb}}}(\tau_1) = G_{\mathrm{PD2}} S_{\mathrm{PD2}} \left[\frac{1}{\tau_{\mathrm{rep}}} \int_{-\infty}^{+\infty} A_{\mathrm{prp}}(z_{17}, \bar{t} - \tau_1) \, A_{\mathrm{pmp}}^{*}(z_{17}, \bar{t}) \, d\bar{t} \right] e^{-j(\omega_c + \omega_{\mathrm{prb}})\tau_1} e^{j(\varphi_{\mathrm{prb}} - \varphi_{\mathrm{pmp}})} .$$

$$\tag{2.29}$$

The magnitude of $U_{\mathrm{PD2},\omega_{\mathrm{rep}}}(\tau_1)$ thus gives directly the magnitude of the cross-correlation between the pump and the probe impulse. The maximum of $U_{\mathrm{PD2},\omega_{\mathrm{rep}}}(\tau_1)$ reveals the position of TS1 for which the centers of the probe and the pump impulse coincide. The width of $U_{\mathrm{PD2},\omega_{\mathrm{rep}}}(\tau_1)$ gives an estimate for the range over which the pump and the probe impulses overlap.

Chapter 3

Understanding Waveguide Loss: Surface Roughness in High Index-Contrast Waveguides

3.1 Introduction

High index-contrast optical waveguides are key elements in the development and design of nanophotonic devices. Although these waveguides provide excellent confinement of the light to the waveguide core, their high index-contrasts make them particularly sensitive to sidewall roughness, which will be present even for sophisticated fabrication processes. Sidewall roughness usually arises due to imperfections in the mask used to etch the waveguide as well as to anisotropies in the etching process itself, and can result in attenuation due to radiative scattering into the surrounding material and into backward-propagating modes. Attenuations have been measured to be 0.24 dB/mm for strip-like silicon-on-insulator (SOI) waveguides [24], for square waveguides the attenuation can be considerably larger (around 1.3 dB/mm) [111], and measurements of InGaAs/InP pedestal waveguides have revealed attenuations as large as 4 dB/mm [26]. A study on losses in Silicon waveguides can be found in [116], in which the authors attributed the low measured attenuations to the smoothness of the waveguides fabricated. Although the magnitude of the roughness-induced loss can be reduced by optimizing the fabrication process [116, 111], this attenuation is considered an unavoidable feature of high index-contrast waveguides, and must be taken into account when designing waveguide-based functional elements. It is therefore crucial to be able to quantitatively predict the magnitude of the roughness-induced losses.

Purely numerical methods are unsuited for the purpose of studying sidewall roughness because they require an extremely fine discretization in order to resolve the sidewall perturbation. In addition, they give little insight into the physics of the loss-mechanisms, and so are not appropriate for producing design rules for the creation of low-loss waveguides. By contrast, semi-analytical methods which are based on the fact that the perturbation is small are potentially able to give both quantitative predictions and physical understanding in precisely those situations for which conventional brute-force computer algorithms become unwieldy or collapse. In particular, coupled mode theory [72] describes the fields in terms of the guided and radiation modes of an unperturbed waveguide. A set of equations which gives the coupling between these various modes can then be derived, and, if the sidewall deviation is small, this method can be used to semi-analytically predict the amount of energy reflected and radiated. As long as both the guided and radiation modes can be calculated, coupled mode theory is straightforward to apply and leads to a good understanding of the physical mechanisms causing the attenuation. The guided modes of most waveguides can be calculated fairly easily (although the problem of finding accurate solutions is still a non-trivial task — see [102] for a good overview) however the radiation modes have so far only been constructed in relatively simple situations, namely, for two-dimensional (2D) slab waveguides [72], for free-space [74], and for optical fibres [98]. Modern integrated optical waveguides typically have a square or rectangular cross-section, and in addition are constructed out of high index-contrast materials such as silicon. For these structures it is expected that the 2D, low-contrast, or circularly symmetric radiation modes previously applied cannot be relied on for quantitative estimates of the attenuation. In order to apply the coupled-mode approach to relevant problems in integrated optics the fully 3D radiation modes which are appropriate for rectangular high-contrast waveguides need to be constructed.

The first theories specifically applied to roughness in integrated optical waveguides were developed in the 1960s and 1970s by Marcuse [70, 71, 73], and were later presented in complete form in his two well-known books [72, 74]. In these papers, the coupled mode formalism was written down for an arbitrary three-dimensional waveguide, however the three-dimensional (3D) radiation modes for a rectangular waveguide were not available at that time, and so results were presented for sidewall perturbations of 2D slab waveguides and for optical waveguides with circular cross-sections. Using a simple model of specular reflection inside a thin film, Tien [108] was able to derive a simple formula for the scattering losses of modes guided within these structures. The 2D coupled-mode formalism was continued by Payne and Lacey [84] and has more recently been applied to three-dimensional rectangular waveguides, where the 3D structure was replaced by a 2D one using an effective index method [64, 41]. The radiation modes of the 3D waveguide were then assumed to be approximated by 2D planar radiation modes. One of these studies [64] achieved good agreement between this numerical model and experiments on strip-like waveguides, which are in any case quite well approximated by a 2D slab. Although such an effective index approximation may work well for high aspect-ratio guides, the 3D radiation modes possess an extra degree of freedom in comparison to the 2D modes, and it is expected that a model which could include this extra dimensionality would yield more accurate results.

A fully 3D and increasingly popular approach for calculating radiation loss in photonic waveguides is known as the volume-current method [60], in which the waveguide perturbation is represented as a polarization volume current density. The loss from the waveguide can then be calculated by integrating the out-going power of the resulting radiation field. However, as has recently been noted by Johnson *et al.* [50], the classical application of this method is only accurate for low index-contrast waveguides and may thus be incorrect by as much as an order of magnitude for strongly guiding structures. These authors have published a correction to this method for high index-contrast materials, and have examined the role played by a surrounding photonic crystal in reducing the radiation loss [50].

In this study the loss due to random wall perturbation of a high index-contrast waveguide was calculated by summing the scattered powers incoherently along the waveguide, and the study of correlated disorder was left to future work.

Other researchers have managed to extend the coupled mode theory to three dimensions for specific situations. It was noted by Snyder *et al.* that Marcuse's earlier expressions for the 3D radiation modes of round waveguides were not orthogonal, and the corrected versions were published in [98]. The authors used these modes to study scattering out of optical fibres with random isotropic variations in radius [97]. Although these authors studied only weakly guiding materials, the expressions which were derived could as well have been applied to high-index, albeit round, structures. More recently, other authors [65] have also used this theory to study radiation from low index-contrast square waveguides, by assuming that the radiation modes of the guide are well approximated by their 3D free-space counterparts. While all of the above-mentioned studies work well for their intended applications, the waveguides previously considered have been either two-dimensional, low-contrast, or circularly symmetric. A key extension for integrated optics is the calculation of 3D radiation modes for the high index-contrast rectangular waveguides which are of practical importance.

In this chapter, we present a semi-analytical model that gives insight into the physical processes involved in waveguide attenuation and which permits us to derive design guidelines for low-loss integrated optical waveguides. The method was developed by Poulton *et al.* and has been published in [86]. It is based on a version of the coupled mode theory developed by Marcuse [72, 74], adapted to include the 3D radiation modes of a high-contrast rectangular waveguide. The construction of these modes is performed using an adaptation of the semi-analytical method of Goell [39], in which the fields inside and outside the waveguide are expanded in terms of cylindrical harmonics and are then matched at the boundary. The radiation modes constructed here are of high interest themselves, and could be used to study waveguide bends and tapers, attenuation from high-contrast Bragg waveguides, or to study efficient coupling into ultra-compact optical devices. We apply these modes to the study of roughness-induced attenuation of high-index contrast rectangular waveguides. It is found that the power lost to radiation is much larger than that reflected into the backward-propagating guided mode, and that there exists a stark contrast between the attenuations of the TE and TM guided modes. We also find good agreement between this semi-analytical method and direct computation of the attenuation using a finite-difference time domain (FDTD) algorithm, whereas the 2D effective index model is found to give results which can be relied upon only qualitatively. We note that a comparison with the volume-current method given in [50] is not feasible because in this study the sidewall roughness was assumed to be uncorrelated. The general applicability of the semi-analytical method presented here gives considerable physical insight into the processes involved in waveguide attenuation, and results in guidelines which can be used for the design of integrated optical devices.

It is important to compare the predictions made by the theoretical model with the attenuation measured from a real device. However, in order to perform such a comparison, one must first determine the physical parameters of sidewall roughness that occur in practice. To this end we present atomic force microscopy measurements of the surface topology of a high index-contrast waveguide: specifically, an InGaAsP/InP pedestal waveguide, fabricated using chlorine-based chemical-assisted ion-beam etching (CAIBE). We find that the model of sidewall roughness with an exponentially decaying autocorrelation function fits well, although the surface structure of the measured waveguide sidewalls exhibits an additional periodicity. The measured roughness parameters are then used in the 3D semi-analytical model and compared with the measured attenuation of the waveguide in order to check the model's validity.

This chapter is divided into 5 parts. In Section 3.2 we adopt the mode coupling formalism developed in Chapter 1. In Section 3.3 we show how to construct the radiation mode fields. In Section 3.4 we examine the effects of sidewall roughness on a square waveguide, and discuss the dominant physical mechanisms behind waveguide attenuation. In Section 3.5, the results are then compared with those from other methods, including predictions from the 2D effective index method and direct calculation using a finite-difference time-domain (FDTD) algorithm. Finally, in Section 3.6 we present the experimentally measured characteristics of the sidewall roughness of the fabricated waveguide, and compare the measured attenuation of this waveguide with that predicted by the semi-analytical model.

Parts of the following chapter have been published in a journal article [J2], see pp. 199.

3.2   Coupled Mode Theory

We consider the waveguide geometry as shown in Figure 3.1. For the analysis which follows we consider the waveguide to be oriented with its center lying on the z-axis, and possessing a cross section lying in the (x, y) plane which is a perturbation on an "ideal" rectangular waveguide of width $2w$ and height $2d$. We consider the perturbation to be constant in y, so that the waveguide edge is described by the coordinate $x = w + f(z)$. The waveguide is surrounded by vacuum (i.e. we do not take into account the effects of a substrate) with permittivity ϵ_0 and permeability μ_0, and the waveguide itself has permittivity ϵ_1 and is assumed to be non-magnetic, and so has the same permeability as the surrounding free-space. We denote the refractive index of the waveguide as $n_g = \sqrt{\epsilon_1/\epsilon_0}$, and thus the refractive index as a function of position can be written

$$n(x, y, z) = \begin{cases} n_g & \text{for } -w \leq x \leq w + f(z), \ -d < y \leq d, \\ 1 & \text{otherwise}. \end{cases} \tag{3.1}$$

We denote the refractive index of the ideal waveguide by $n_0(x, y)$, where

$$n_0(x, y) = \begin{cases} n_g & \text{for } |x| \leq w, \ |y| \leq d, \\ 1 & \text{otherwise}. \end{cases} \tag{3.2}$$

For the moment it is sufficient to consider the waveguide perturbation to exist only on a single side of the waveguide.

Mode Coupling Equations

The sidewall perturbations of the waveguide couple the guided and radiation modes according to Eqs. (1.64) and (1.65), where $\Delta\epsilon_r(\mathbf{r}) = \epsilon_0 \left(n^2(x, y, z) - n_0^2(x, y) \right)$. In the follow-

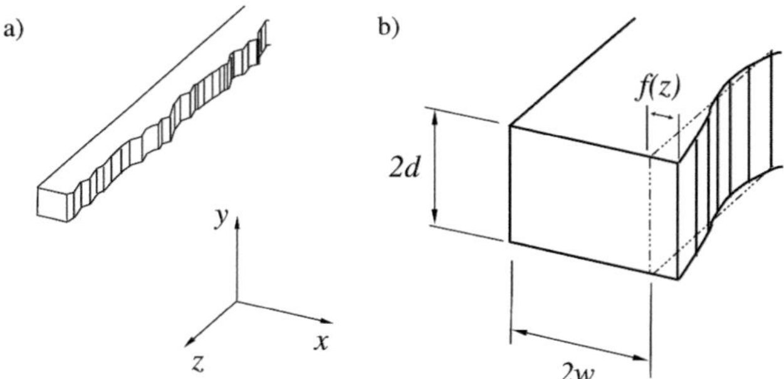

Fig. 3.1. Geometry of the waveguide with rough sidewall. a) Sketch of the waveguide; b) Sectioned view, showing the parameters used in the analysis. The waveguide is centered about the z-axis.

ing, we are not interested in the dynamics of the envelope $A_{\zeta,\beta_c}(z,t,\omega_c)$, and we consider only one carrier frequency ω_c. The time dependence and the frequency argument of $A_{\zeta,\beta_c}(z,t,\omega_c)$ are therefore omitted. Further, to evaluate Eq. (1.64) numerically, the integral must be discretized. Merging the modal subscripts (ζ,β_c) and (ζ',β_c') into single subscripts μ and ν, the combined integration and summation $\sum \int$ can be replaced by a simple sum $\sum_\mu$. The quantity $A_{\zeta,\beta_c}(z)\,\mathrm{d}\beta_c$ in Eq. (1.64) is then replaced by a mode amplitude $c_\mu(z)$. Finally, for the remainder of this chapter, bold symbols $\mathbf{E}_\mu(x,y)$ and $\mathbf{H}_\mu(x,y)$ are used for the vectorial electric and magnetic mode fields $\mathcal{E}_{\zeta,\beta_c}(x,y,\omega_c)$ and $\mathcal{H}_{\zeta,\beta_c}(x,y,\omega_c)$. All mode fields are normalized to the same power $P = \mathcal{P}_{\zeta',\beta_c'} = \mathcal{P}_{\zeta,\beta_c}$.

Eqs. (1.64) and (1.65) can thus be rewritten in a simpler form,

$$\frac{\mathrm{d}c_\nu(z)}{\mathrm{d}z} = -\mathrm{j}\sum_\mu K_{\nu\mu}(z)\exp[-\mathrm{j}(\beta_\mu - \beta_\nu)z]\,c_\mu(z)\,, \tag{3.3}$$

The coupling between the μth and ν^{th} modes is given by

$$K_{\nu\mu}(z) = \frac{\omega\epsilon_0}{4P}\int_{-\infty}^{+\infty}\mathrm{d}x\int_{-\infty}^{+\infty}\mathrm{d}y\,\left\{(n^2 - n_0^2)\,\mathbf{E}_\nu^*\cdot\mathbf{E}_\mu\right\} \tag{3.4}$$

The coupling between the modes is then directly dependent on the index contrast of the waveguide via the term $(n^2 - n_0^2)$, as well as on the values of the μth and νth modes in the vicinity of the perturbation. In the following, μ refers to a guided mode, the power attenuation of which is investigated. The ensemble of guided modes, into which the power is transferred, is denoted by a subscript ν.

Perturbation Solution

The coupled amplitude Eqs. (3.4) form a complicated linear system which governs the interaction between the modes caused by the waveguide perturbation. It is difficult to solve this system without a suitable approximation. If the magnitude of the waveguide perturbation is small, then it is a valid assumption that, over a short distance in z, the mode-amplitude of the excited guided mode will remain large in comparison to all the other waveguide modes. In this case we can integrate Eq. (3.3) directly, and thereby obtain the amplitude of any given mode at the position $z = L$. If, at $z = 0$, the incident mode-amplitude is $c_\mu(0)$ and all the other mode-amplitudes are equal to zero, then after a short distance L the amplitude of the νth mode is

$$c_\nu(L) = -\mathrm{j}\,c_\mu(0) \int_0^L K_{\nu\mu}(z)\exp[-\mathrm{j}(\beta_\mu - \beta_\nu)z]\mathrm{d}z\,. \tag{3.5}$$

It remains to find simple expressions for the coupling terms $K_{\mu\nu}(z)$, which have a non-trivial dependence on the perturbation function $f(z)$. The expression under the integral appearing in Eq. (3.4) is non-zero only in the narrow strip in which the perturbed waveguide differs from the unperturbed guide, because elsewhere the term $(n^2 - n_0^2)$ vanishes. In accordance with our earlier assumption that the perturbation is small, we can approximate the fields appearing under the integral in (3.4) by their values close to the waveguide boundary. Some of the field components are discontinuous over the waveguide boundary (as is the refractive index) and so for high-contrast materials care must be taken to evaluate the fields on the correct side. In this we depart from the treatment given by Marcuse[72], who treats the field components on both sides of the waveguide boundary as being approximately equal. If the function $f(z) > 0$, then, representing the term under the integral by the function $g(x, y)$ and noting that $g(x, y) \neq 0$ only when $w \leq x \leq w + f(z)$, $|y| \leq d$, we can make the approximation

$$K_{\nu\mu}(z) = \frac{\omega\epsilon_0}{4P}\int_w^{w+f(z)} \mathrm{d}x \int_{-d}^d \mathrm{d}y\, g(x, y) \approx \frac{\omega\epsilon_0}{4P}f(z)\int_{-d}^d \mathrm{d}y\,[g(x, y)]_{x=w+\bar\sigma/2}\,. \tag{3.6}$$

The resulting expression in the square brackets is evaluated at half the sidewall root-mean-square RMS deviation $\bar\sigma$ outside the unperturbed boundary. This approximation is based on the assumption that the mode fields vary linearly very close to the unperturbed waveguide edge. Therefore, if the function $f(z) > 0$, the coupling terms $K_{\mu\nu}$ can be approximated by

$$K_{\mu\nu}(z) = f(z)\widehat{K}_{\mu\nu}^{(e)} \tag{3.7}$$

where

$$\widehat{K}_{\nu\mu}^{(e)} = \frac{\omega\epsilon_0}{4P}(n_g^2 - 1)\int_{-d}^d \mathrm{d}y\,\left[\mathbf{E}_\nu^* \cdot \mathbf{E}_\mu\right]_{x=w+\bar\sigma/2} \tag{3.8}$$

For the case $f(z) < 0$, the discontinuities in the field components leads to coupling which is in general different from the coupling for the case $f(z) > 0$. The coupling term arising from the fields inside the waveguide can be written

$$K_{\mu\nu}(z) = f(z)\widehat{K}_{\mu\nu}^{(i)}\,, \tag{3.9}$$

where a procedure analogous to the one leading to Eq. (3.8) yields

$$\widehat{K}_{\nu\mu}^{(i)} = \frac{\omega\epsilon_0}{4P}(n_g^2 - 1)\int_{-d}^d \mathrm{d}y\,\left[\mathbf{E}_\nu^* \cdot \mathbf{E}_\mu\right]_{x=w-\bar\sigma/2} \tag{3.10}$$

The general form of the z-dependence of the coupling terms for small waveguide perturbations is then

$$K_{\nu\mu}(z) = f(z)\widehat{K}_{\nu\mu}^{(i,e)}\,. \tag{3.11}$$

Substituting (3.11) into (3.5), the amplitude of the νth mode after a propagation distance L (under the assumption that L is small enough that the incident mode remains approximately undepleted) is

$$c_\nu(L) = -\mathrm{j}\,\sqrt{L}\,c_\mu(0)F(\beta_\mu - \beta_\nu)\widehat{K}_{\nu\mu}^{(i,e)}\,, \tag{3.12}$$

where

$$F(\beta_\mu - \beta_\nu) = \frac{1}{\sqrt{L}}\int_0^L f(z)\exp[-\mathrm{j}(\beta_\mu - \beta_\nu)z]\mathrm{d}z \tag{3.13}$$

may be interpreted as a Fourier spectrum of the truncated sidewall perturbation function $f(z)$. The coupling from the μth to the νth mode is then directly dependent on the Fourier spectrum of the perturbation function $f(z)$. If the perturbation function has a large spectral amplitude F at a spatial frequency which "makes up the difference" between the propagation constants of the two modes, then the coupling from the (guided) μ^{th} mode to the νth (radiation) mode will be enhanced. For a purely sinusoidal sidewall deformation, the μth mode will radiate only into modes with a single propagation constant which is determined by the period of the perturbation.

The coupling between the different modes also depends *indirectly* on the propagation constants via the coupling terms $\widehat{K}_{\nu\mu}^{(i,e)}$, which are also functions of β_ν and β_μ since the modal fields will change for different propagation constants. The fields near the boundaries are then essential for the quantitative calculation of the coupling from one mode to the other. The fields of guided modes have been extensively studied and calculated for many years [102, 114]. It is the purpose of the next section to find expressions for the fields of the waveguide radiation modes.

3.3 Construction of Radiation Modes

We proceed by adapting the semi-analytical method developed by Goell [39] to the case of radiation modes. Although this method is designed for constructing guided modes, it can be extended to calculate the response of a high-contrast waveguide to a specified incoming wave. The combined field, including incoming wave and response, forms an individual radiation mode. The method chosen for the construction has several advantages for high index-contrast waveguides: Firstly, the accuracy with which the modes can be constructed is not dependent on the condition that the mode is weakly guided. Secondly, the expansion in terms of cylindrical harmonics is suitable for the application to compact waveguides, since the dominant energy transport away from the waveguide will be simply expressed in the form of outgoing waves. In addition, for any localized waveguide the terms in the field expansion will decay exponentially, thus ensuring high accuracy. Finally it will be seen that the orthogonality and normalization of these modes is relatively straightforward.

One subtlety of the modal expansion given in Eqs. (1.35) and (1.36) is that the total field scattered by an imperfection of the waveguide must consist only of outgoing waves in the far field, whereas the radiation modes consist of both incoming and outgoing waves. This would appear to make an expansion of a scattered wave in terms of these radiation modes problematic. However, it has been observed[72, p. 21] that because the radiation modes form a continuum, it is impossible for a scattering source to excite a single radiation mode without simultaneously exciting an infinite number of radiation modes in the spectral vicinity. The outgoing parts of the radiation modes then add coherently, while the incoming parts fail to interfere constructively and are entirely cancelled.

It can be shown (see Appendix D.1) that the radiation modes are orthogonal provided the incoming waves chosen in the construction are themselves orthogonal. The question of completeness is more involved. For the purpose of calculating attenuation loss due to imperfections in the waveguide sidewall, we need consider only those radiation modes which are able to carry energy away from the waveguide. These are the modes for which the transverse propagation constant ρ remains positive and real in both regions of the guide. This places a limitation on the values of propagation constant; if we consider the radiation mode with mode index ν, then β_ν must lie within the range

$$-k \leq \beta_\nu \leq k \,. \tag{3.14}$$

This restriction has consequences regarding the completeness of the mode expansion — the modes which are evanescent in the z-direction are necessary for the reconstruction of the total fields in the near vicinity of the perturbed waveguide boundary, and so in this region the set of modes is no longer strictly complete. However, these modes are not able to transport energy away from the guide, and so can be safely neglected for the purposes of calculating scattering losses.

The strategy is now as follows: the radiation modes are defined as resulting from the response of the waveguide to an incident field possessing a given symmetry and polarization. The longitudinal or z-components of each field can be written as a series of cylindrical harmonic functions both internal and external to the waveguide, and the tangential components (i.e., those which lie in the (x, y) plane and are tangential to the waveguide surface) follow from the longitudinal components [101, p. 361]. The unknown coefficients in the harmonic expansion are determined by matching the internal and external fields of both tangential and longitudinal components at a number of specified points on the waveguide boundary.

In order to construct the modes in terms of cylindrical harmonic functions, we now write the spatial dependence of the vector field components in terms of cylindrical coordinates (r, θ, z), as shown in Figure 3.2. The mode fields $(\mathbf{E}_\nu, \mathbf{H}_\nu)$ of the νth radiation mode obey the Helmholtz equations

$$\begin{aligned}
\left(\nabla_t^2 + n_0^2(x, y)k^2 - \beta_\mu^2\right) \mathbf{E}_\mu(x, y) &= 0 \\
\left(\nabla_t^2 + n_0^2(x, y)k^2 - \beta_\mu^2\right) \mathbf{H}_\mu(x, y) &= 0 \,,
\end{aligned} \tag{3.15}$$

where we have introduced the transverse Laplacian operator $\nabla_t^2 = \partial^2/\partial x^2 + \partial^2/\partial y^2$. Making use of the fact that $\rho = (k^2 - \beta_\nu^2)^{1/2}$, and defining the transverse wavenumber in the internal region of the waveguide as $\eta = (n_g^2 k^2 - \beta_\nu^2)^{1/2}$, the z, or longitudinal, components of the electric and magnetic fields can be written

$$E_L^i(r,\theta) = [\mathbf{E}_\nu^i]_z = \sum_{\ell=0}^{\infty} A_\ell J_\ell(\eta r) \sin(\ell\theta + \varphi), \tag{3.16}$$

$$H_L^i(r,\theta) = [\mathbf{H}_\nu^i]_z = \sum_{\ell=0}^{\infty} B_\ell J_\ell(\eta r) \cos(\ell\theta + \varphi), \tag{3.17}$$

in the internal region, and

$$E_L^e(r,\theta) = [\mathbf{E}_\nu^e]_z = \sum_{\ell=0}^{\infty} C_\ell H_\ell^{(1)}(\rho r) \sin(\ell\theta + \varphi) + P_N J_N(\rho r) \sin(N\theta + \varphi), \tag{3.18}$$

$$H_L^e(r,\theta) = [\mathbf{H}_\nu^e]_z = \sum_{\ell=0}^{\infty} D_\ell H_\ell^{(1)}(\rho r) \cos(\ell\theta + \varphi) + Q_N J_N(\rho r) \cos(N\theta + \varphi). \tag{3.19}$$

in the external region. Here we have written the longitudinal components as E_L and H_L rather than $[\mathbf{E}_\nu^i]_z$ and $[\mathbf{H}_\nu^i]_z$ for the sake of notational simplicity. The terms J_ℓ and $H_\ell^{(1)}$ are the ℓth order Bessel functions and Hankel functions of the first kind respectively. The multipole coefficients P_N and Q_N determine the incident field, and thus the symmetry of the radiation mode. The multipole coefficients A_ℓ, B_ℓ, C_ℓ, D_ℓ must be determined in order to reconstruct the mode completely.

In the above equations, P_N and Q_N represent the magnitudes of waves incident on the waveguide, whereas the coefficients C_ℓ and D_ℓ give the magnitudes of the outgoing (or "response") waves. If $P_N = 0$, $Q_N \neq 0$, the incident wave is a pure "H-wave" — i.e., the z component of the electric field is zero at infinity whereas the z component of the magnetic field remains finite. If $P_N \neq 0$, $Q_N = 0$ then the reverse situation applies, and the incident wave is a pure "E-wave". A given radiation mode should possess only one of these two polarizations, and so in the following we assume without loss of generality that $P_N Q_N = 0$. The quantity N gives the far-field multipole order of the radiation mode and so represents a symmetry in the angular coordinate, albeit at infinity; in the vicinity of the origin, the fields will be distorted by the presence of the waveguide. We also note that the waveguide is symmetric about the x axis and so the fields must be either symmetric or antisymmetric about this axis. These two possible additional symmetries are represented by the phase factor φ, which is equal either to 0 or to $\pi/2$. In order to keep track of these various symmetries, to each radiation mode we assign the label ζ, which we define to be

$$\zeta = (N, \varphi, \Pi), \tag{3.20}$$

where Π labels the incident field polarization: either $\Pi =$ 'E' for an incident E-wave or $\Pi =$ 'H' for an incident H-wave. The label ζ, together with the propagation constant β_ν, is sufficient to identify an individual radiation mode.

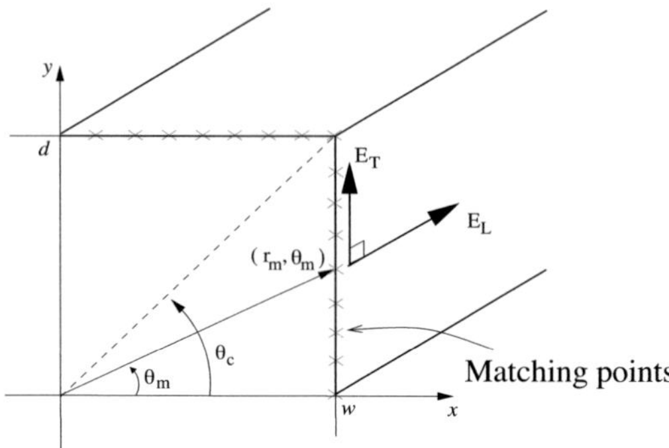

Fig. 3.2. Section of the first quadrant of the ideal waveguide, showing the matching points (r_m, θ_m) and the longitudinal and tangential electric field vectors.

The field components perpendicular to the waveguide axis can be reconstructed from the z components of the electric and magnetic fields. In cylindrical polar coordinates, the relations are [39]

$$E_r = \frac{-j\beta}{n_0^2 k^2 - \beta^2} \left(\frac{\partial E_z}{\partial r} + \frac{\mu_0 \omega}{\beta r} \frac{\partial H_z}{\partial \theta} \right) \tag{3.21}$$

$$H_r = \frac{-j\beta}{n_0^2 k^2 - \beta^2} \left(-\frac{n^2 k}{Z_0 \beta r} \frac{\partial E_z}{\partial \theta} + \frac{\partial H_z}{\partial r} \right) \tag{3.22}$$

$$E_\theta = \frac{-j\beta}{n_0^2 k^2 - \beta^2} \left(\frac{1}{r} \frac{\partial E_z}{\partial \theta} - \frac{\mu_0 \omega}{\beta} \frac{\partial H_z}{\partial r} \right) \tag{3.23}$$

$$H_\theta = \frac{-j\beta}{n_0^2 k^2 - \beta^2} \left(\frac{n^2 k}{Z_0 \beta} \frac{\partial E_z}{\partial r} + \frac{1}{r} \frac{\partial H_z}{\partial \theta} \right) , \tag{3.24}$$

where we have introduced the impedance of free space $Z_0 = \sqrt{\mu_0/\epsilon_0}$, and the field components can represent either the internal or external fields.

The tangential components of both the electric and magnetic fields must be continuous across the waveguide boundary. For a rectangular waveguide, we write the tangential components which lie in the (x, y) plane as

$$E_T(r, \theta) = E_r(r, \theta)R(\theta) + E_\theta(r, \theta)T(\theta) \tag{3.25}$$

$$H_T(r, \theta) = H_r(r, \theta)R(\theta) + H_\theta(r, \theta)T(\theta) , \tag{3.26}$$

where the functions $R(\theta)$ and $T(\theta)$ incorporate the geometry of the waveguide boundary. If $\theta = \theta_c$ represents the angle of the waveguide corner (see Figure 3.2) then

$$R(\theta) = \begin{cases} \sin\theta & \theta < \theta_c \\ \cos\theta & \theta \geq \theta_c \end{cases} \tag{3.27}$$

$$T(\theta) = \begin{cases} \cos\theta & \theta < \theta_c \\ -\sin\theta & \theta \geq \theta_c \,. \end{cases} \tag{3.28}$$

Here we have studied a rectangular waveguide; the extension for waveguides with more general cross-sections simply involves replacing the sinusoidal functions $R(\theta)$ and $T(\theta)$ with more general expressions.

Field matching at waveguide boundaries

The field components E_T, H_T, E_L and H_L must be continuous at the waveguide boundaries, and can thus be matched at each of the points (r_m, θ_m). Due to the symmetry of the fields, we are able to choose matching points which lie entirely within the first quadrant of the waveguide (as shown in Figure 3.2). We choose M matching points, located at positions (r_m, θ_m), where $\theta_m = (m-1)\pi/2(M-1)$; $m = 1...M$. The positions r_m then follow from the position of the waveguide edge in polar coordinates. The number of matching points, together with the symmetry of the incident field and the phase factor φ, then determines the number of harmonics which are necessary to reconstruct the fields.

In order to reconstruct the fields with sufficient accuracy, the multipole series in Eqs. (3.17) and (3.19) must be appropriately truncated, and care must be taken that the problem is neither over- nor under-specified, and that the multipole coefficients chosen match the symmetry of the incident field. In particular, the E-waves (H-waves) feature the following symmetry properties: For the case in which N is odd and $\varphi = 0$, the z-component of the incident electric (magnetic) field is symmetric (antisymmetric) about the y-axis, and so only odd multipole coefficients appear in the series expansion. In this case we choose the multipole coefficients to lie in the range $1, 3, ...2M - 1$. For the case in which N is even and greater than zero and $\varphi = 0$, the z-component of the incident electric (magnetic) field is antisymmetric (symmetric) about the y-axis, and the multipole coefficients should be even and can be chosen to lie in the range $0, 2, ...2M - 2$. For $\varphi = \pi/2$, the symmetry properties of the incident fields are simply reversed. The special case $N = 0$ retains the features both of even and odd symmetry. If $\varphi = 0$ then symmetry dictates that $P_N = 0$; alternatively if $\varphi = \pi/2$ then $Q_N = 0$. In both cases we choose even multipole coefficients in the range $0, 2, ...2M - 2$.

The matching equations for the longitudinal components of the fields can be written in matrix form

$$E^{LA}A = E^{LC}C + E^{LP}P \tag{3.29}$$

$$H^{LB}B = H^{LD}D + H^{LQ}Q \,, \tag{3.30}$$

where we have gathered the multipole coefficients into column vectors: A, B, C and D are M-element column vectors consisting of coefficients A_ℓ, B_ℓ, C_ℓ and D_ℓ respectively, and P and Q are the M-element column vectors consisting of the quantities $P_N\delta_{\ell,N}$ and $Q_N\delta_{\ell,N}$. The $M \times M$ matrices E^{LA}, E^{LC}, H^{LB}, and H^{LD} are somewhat cumbersome to derive, and are given in Appendix D.2. For the transverse fields the matching equations are given by

$$E^{TA}A + E^{TB}B = E^{TC}C + E^{TD}D + E^{TP}P + E^{TQ}Q \tag{3.31}$$

$$H^{TA}A + H^{TB}B = H^{TC}C + H^{TD}D + H^{TP}P + H^{TQ}Q \,. \tag{3.32}$$

The matching equations can now be represented as the $4M \times 4M$ matrix system

$$\begin{pmatrix} E^{LA} & 0 & -E^{LC} & 0 \\ 0 & H^{LB} & 0 & -H^{LD} \\ E^{TA} & E^{TB} & -E^{TC} & -E^{TD} \\ H^{TA} & H^{TB} & -H^{TC} & -H^{TD} \end{pmatrix} \begin{pmatrix} A \\ B \\ C \\ D \end{pmatrix} = \begin{pmatrix} E^{LP}P \\ H^{LQ}Q \\ E^{TP}P + E^{TQ}Q \\ H^{TP}P + H^{TQ}Q \end{pmatrix} \tag{3.33}$$

The terms on the right-hand side of Eq. (3.33) come from the incident field, and so depend on the choice of the coefficients P_N and Q_N. Given a choice of polarization/symmetry ζ and a propagation constant β_ν, the matrix system (3.33) is readily inverted in order to find the unknown coefficients A_ℓ, B_ℓ, C_ℓ and D_ℓ.

Examples of Radiation Modes

We present a number of the radiation modes that have been calculated by solving the matrix system (3.33) and substituting the coefficients A_ℓ, B_ℓ, C_ℓ, D_ℓ, P_N and Q_N into the equations for the fields (3.17) and (3.19). For the calculations we chose a square silicon waveguide ($n_g = 3.48$) with a width and height of 0.4μm, at a frequency of 194 THz corresponding to a free-space propagation constant $k = 4.06\,\mu\text{m}^{-1}$. The radiation mode propagation constant was chosen to be $\beta_\rho = 1.964\,\mu\text{m}^{-1}$. In order to ensure accuracy of the calculation we chose $M = 23$ matching points, thus the maximum multipole order occurring in the field expansions was either 45 or 46, depending on the symmetry of the fields.

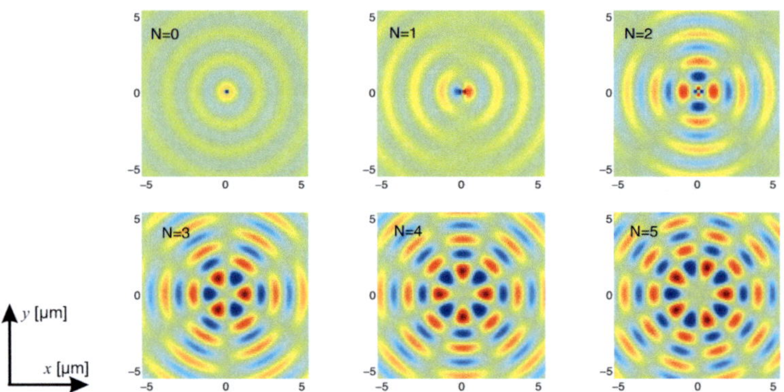

Fig. 3.3. Radiation modes for a square silicon waveguide. The real part of the z-component of the magnetic field is shown. All modes have $\varphi = 0$, the incident wave is an H-wave, and the far-field multipole orders are shown for the values $N = 0...5$. The waveguide width and height are $2w = 2d = 0.4\,\mu$m, its operating frequency is $f = 194\,\text{THz}$. The units of all axes are in micrometers.

Figure 3.3 shows the z-component of the real part of the magnetic field for $\varphi = 0$, 'H-wave' incident field, with N taking values in the range 0...5. It is noted that while the high-order radiation modes are similar to their free-space counterparts given in [97], the lower-order modes differ considerably in the region of the waveguide. This is important because it is the quantitative value of the radiation mode on the waveguide edge which determines the coupling. The difference can be seen explicitly in Figure 3.4, where we compare H_z components of the radiation modes in the vicinity of the waveguide for the high-index contrast waveguide with the free-space radiation modes derived in [97]. It can be seen that the presence of the waveguide significantly distorts the modes, and so free-space radiation modes can only be used as an approximation for low index contrasts.

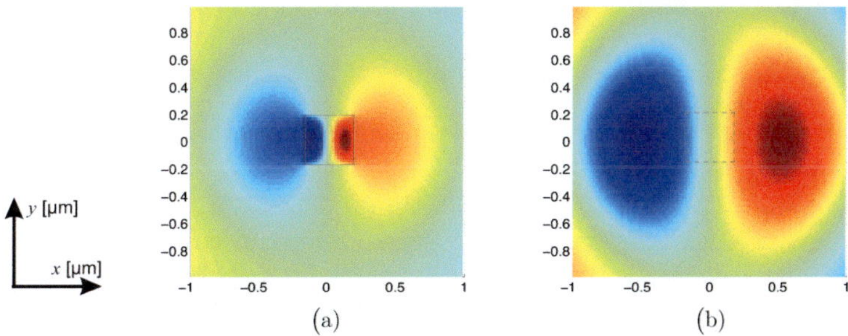

Fig. 3.4. (a) Close up in the vicinity of the waveguide of the radiation mode corresponding to $N = 1$ in Figure 3.3. (b) The same radiation mode in the absence of a waveguide — i.e. the free-space radiation mode. In both cases the units are in micrometers. It can be seen that the presence of the waveguide significantly distorts the field values in the vicinity of the origin, and so using such free-space radiation modes will lead to inaccuracies when predicting the attenuation.

3.4   Loss Due to Sidewall Roughness

We now examine the specific case where a random wall perturbation causes coupling from a guided mode to the radiation modes, and thereby power loss in the guide. At a given position z, the power carried by any mode labelled by the subscript ν is $P|c_\nu(z)|^2$. If ν represents a radiation mode, then the total power lost from the incident guided mode into the radiation field over a distance L is then just the sum of all the powers in the individual radiation modes, and is equal to

$$|\Delta P| = \sum_\nu \int_{-k}^{k} |P||c_\nu(L)|^2 \frac{|\beta_\nu|}{\rho} \mathrm{d}\beta_\nu \, . \tag{3.34}$$

In this expression we have explicitly separated the integral over ρ from the sum over ν, and the remaining primed sum extends over the radiation mode symmetries given by ζ in Eq. (3.20). The integral over ρ has then been replaced by one over β_ν, which enables us to represent the forward and backward propagating waves in a single integral. Note that the limits of the integral reflect the restriction on the β_ν, and thus on our assumption that only "radiating" radiation modes can carry power away from the guide.

Using the expression (3.12), the power attenuation constant for the μth guided mode is then

$$\alpha_\mu \approx 2\frac{|\Delta P|}{P|c_\mu(0)|^2 L} = 2\sum_\nu \int_{-k}^{k} |\widehat{K}_{\nu\mu}^{(i,e)}|^2 \left\langle |F(\beta_\mu - \beta_\nu)|^2 \right\rangle \frac{|\beta_\nu|}{\rho} \mathrm{d}\beta_\nu \ . \tag{3.35}$$

The symbol $\langle \cdots \rangle$ represents an ensemble average over the possible representations of the sidewall deformation functions. The factor 2 on the rhs of Eq. (3.35) comes from the fact that usually both of the waveguide sidewalls are rough, and that both of these sidewalls will contribute to the power loss. If we assume that the roughness on each sidewall is uncorrelated to that on the other, then both sidewalls contribute equally to the attenuation α_μ.

The statistics of this ensemble are determined by the autocorrelation function $R(u_z)$, which is defined as

$$R(u_z) = \langle f(z)f(z - u_z) \rangle \ . \tag{3.36}$$

A commonly used model for waveguide roughness is to use an autocorrelation function of the form

$$R(u_z) = \bar{\sigma}^2 \exp(-|u_z|/D) \ , \tag{3.37}$$

where $\bar{\sigma}$ is the RMS deviation of the perturbation function and D is known as the correlation length. If the integration length L, while being small enough so that the expression (3.12) remains valid, is nevertheless sufficiently longer than the correlation length D then the ensemble-averaged power spectrum occurring in (3.35) can be replaced by the Fourier transform of the autocorrelation function, using the Wiener-Khintchine theorem [83]. For the model (3.37) the result is

$$\left\langle |F(\beta_\mu - \beta_\nu)|^2 \right\rangle = \frac{2\bar{\sigma}^2}{D} \frac{1}{(\beta_\mu - \beta_\nu)^2 + 1/D^2} \ . \tag{3.38}$$

Substitution of this model for the waveguide roughness, together with the expressions for the coupling terms $\widehat{K}_{\mu\nu}^{(i,e)}$, into the expression for the power attenuation coefficient gives two possible expressions for the attenuation: one expression corresponding to those parts of the waveguide for which $f(z) > 0$, and another from the parts of the waveguides for which $f(z) < 0$. It seems reasonable to assume that these two parts will contribute additively to the total loss. This can be justified by noting that the power loss coefficient comes from an integral over z (Eq. (3.5)), to which both positive and negative parts of $f(z)$ contribute equally. We can hence write the power attenuation constant α_μ for a single rough sidewall as

$$\alpha_\mu = \alpha_\mu^{(e)} + \alpha_\mu^{(i)} \ , \tag{3.39}$$

where, for the case $f(z) > 0$, we have

$$\alpha_\mu^{(e)} = 2\left(\frac{\omega\epsilon_0}{4P}\right)^2 (n_g^2 - 1)^2 \sum_\nu \int_{-k}^{+k} \mathrm{d}\beta_\nu \frac{|\beta_\nu|}{\rho} \left\{ \frac{2\bar{\sigma}^2}{D[(\beta_\mu - \beta_\nu)^2 + 1/D^2]} \left| \int_{-d}^{d} \mathrm{d}y \left[\mathbf{E}_{\mu t}^* \cdot \mathbf{E}_{\nu t}\right]_{x=w+\bar{\sigma}/2} \right| \right\}$$
$$\tag{3.40}$$

and for the case $f(z) < 0$ we find

$$\alpha_\mu^{(i)} = 2 \left(\frac{\omega\epsilon_0}{4P}\right)^2 (n_g^2 - 1)^2 \sum_\nu \int_{-k}^{+k} \mathrm{d}\beta_\nu \frac{|\beta_\nu|}{\rho} \left\{ \frac{2\bar\sigma^2}{D[(\beta_\mu - \beta_\nu)^2 + 1/D^2]} \left| \int_{-d}^{d} \mathrm{d}y \left[\mathbf{E}_{\mu t}^* \cdot \mathbf{E}_{\nu t}\right]_{x=w-\bar\sigma/2} \right| \right\}$$

$$(3.41)$$

The fields at $x = w \pm \bar\sigma/2$ are provided by a numerical computation of the guided and radiation modes of the ideal waveguide.

One can immediately make a number of observations as to the dependence of the attenuation on the various waveguide parameters, using Eqs. (3.39-3.41). Apart from the quadratic dependence of the coupling on the RMS roughness $\bar\sigma$, there exists a non-trivial dependence on the correlation length D. As noted in Section 3.2, if the perturbation function possesses a Fourier amplitude which peaks at spatial frequencies close to the difference between the propagation constants of a radiation mode β_ν and the propagating waveguide mode β_μ, then the coupling between the two will be enhanced due to the factor $(\beta_\mu - \beta_\nu)$ on the denominator of Eqs. (3.39-3.41). Because the ensemble-averaged power spectrum is the Fourier transform of the autocorrelation function, the statistics of the random perturbation will play an important role in the behavior of the attenuation.

It can also be seen that there is a direct relation between the magnitudes of the field components of the guided and radiation modes and the attenuation. As mentioned previously, at optical frequencies the transverse electric field terms in the integral will usually dominate; a high index-contrast material will have a jump in the electric field components on the boundary of the waveguide which will also affect the attenuation, and will lead to markedly different behavior for different polarizations.

3.4.1 Results for a Rough Silicon Waveguide

The total loss from the fundamental mode in a square singlemode silicon waveguide with width and height $0.42\,\mu$m and RMS sidewall roughness $\bar\sigma = 5\,$nm, calculated using Eq. (3.39), is shown in Figure 3.5. The attenuation is plotted as a function of the correlation length D. In this calculation we have assumed that the perturbation exists on the two waveguide sidewalls parallel to the y-axis, and that both sidewalls contribute equally to the power attenuation. Both attenuation due to radiation and attenuation arising from reflection into the backward-propagating waveguide mode are taken into account. We examine the two possible polarizations for the fundamental mode — the TE polarization is defined to be that for which the dominant electric field component is parallel to the x-axis and therefore lies perpendicular to the roughened waveguide boundaries, whereas the dominant E-field component of the TM polarization lies parallel to the rough sidewalls. For this calculation, 23 matching points were used in order to accurately reconstruct the radiation fields on the unperturbed waveguide boundaries, and the β_ν integral in Eqs. (3.40) and (3.41) was discretized into 128 points between the values of $-k$ and $+k$. For each value of β_ν there are two types of radiation modes (corresponding to $\phi = 0$ and $\phi = \pi/2$), and all radiation modes in the range from $N = 0$ and $N = 12$ were chosen in the field expansions. Therefore for each possible scattering angle $13 \times 2 = 26$ radiation modes were used to represent the radiation field.

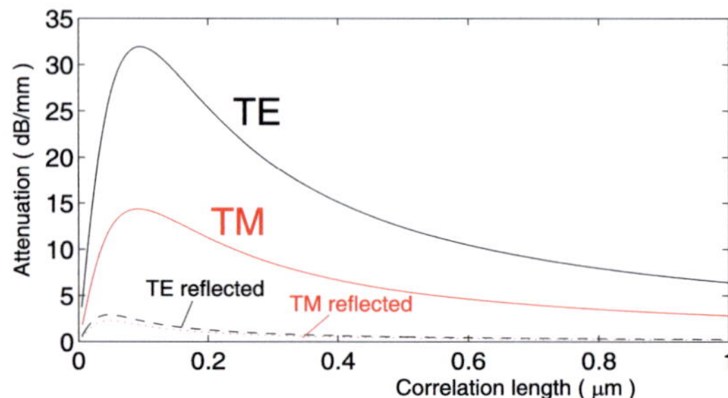

Fig. 3.5. Total attenuation as a function of correlation length D for a square silicon waveguide. It can be seen that TE polarization exhibits a much larger attenuation than TM, and also that the attenuation due to radiative scattering is much larger than that due to reflection. Waveguide dimensions $2w = 2d = 0.42\,\mu$m, RMS sidewall roughness $\bar{\sigma} = 5$ nm, operating frequency $f = 194$ THz

One can see in Figure 3.5 that the dominant loss mechanism for both polarizations is radiation rather than reflection. This is due to the difference in propagation constants, which is larger for the reflected mode than for the entire range of radiation modes. It is also observed that the attenuation of the TE mode is much larger than that of the TM mode. This is to be expected from examination of Eqs. (3.40) and (3.41), since the dominant component of the TE mode undergoes a large jump (proportional to n_g^2) on the waveguide boundary, whereas the dominant component of the TM mode is continuous, and is therefore smaller in the external region of the guide.

An additional feature of the roughness curve is the presence of a maximum, occurring at a correlation length of $D_{\max} = 95$ nm. This behavior has been observed by other studies on waveguide roughness [72, 97, 65], and can be interpreted in the following way: A waveguide with a large correlation length possesses a variation along the propagation direction which is much longer than the effective wavelength of the propagating mode, and hence the scattering is reduced. On the other hand, for short correlation lengths the waveguide boundary varies so rapidly that the waveguide mode cannot resolve the details, and so the attenuation is also reduced. In between these two extremes lies the Mie-scattering regime, where the waveguide irregularities possess the same order of magnitude as the effective wavelength, and in this regime the coupling is maximized.

3.4.2   Design Guidelines

By examination of Eqs. (3.40) and (3.41), one can obtain a rough estimate of the attenuation maximum. The transverse electric field components of the radiation modes are given in Appendix D.1 by Eqs. (D.5d). Assuming that the transverse dimensions of the waveguide are small compared to the free-space wavelength of the light (such that $k\sqrt{w^2 + d^2} < 1$), we can approximate the attenuation for small values of ρr. If we further take into account that $\beta_\mu^2 >> k^2$ for high index-contrast waveguides, the loss as a function of correlation length is approximately proportional to

$$\alpha_\mu(D) \propto \bar\sigma^2 \frac{2kD}{1 - D^2(k^2 - \beta_\mu^2)} \, . \tag{3.42}$$

As to be expected, the RMS roughness $\bar\sigma$ determines the overall losses significantly. However, for a given $\bar\sigma$, a maximum attenuation occurs approximately when

$$D_{\text{max}} \approx \frac{1}{\sqrt{\beta_\mu^2 - k^2}} \, . \tag{3.43}$$

In our case this leads to an approximate maximum of $D_{\text{max}} = 100\,\text{nm}$, which lies very close to the observed value of 95 nm.

The correlation length is a parameter defined by the waveguide fabrication process, and so it is generally easier to modify the propagation constant of the fundamental mode in order to reduce the attenuation. By inspection of (3.43), we can see that this means increasing the propagation constant β_μ as much as is possible. However, the only way of changing the propagation constant of the mode without changing the material is to change the dimensions of the waveguide. A larger waveguide will, in general, possess a larger propagation constant of a fundamental mode for any given frequency. On the other hand, a larger waveguide will also possess a larger waveguide surface which can radiate, and so the attenuation may be increased. The optimal case would be a waveguide which is much wider than it is high, thereby having both a large propagation constant and a small scattering surface. However, care must be taken not to exceed the single-mode limit. This is in fact the design used in the low-loss guides described in [24].

3.5 Comparison with other methods

3.5.1 Comparison with FDTD

In order to lend plausibility to the results given in the previous section we present a comparison with a direct calculation of a number of roughened waveguide realizations using a finite difference time domain (FDTD) algorithm. The calculation of attenuation using such a "brute-force" method is not a simple matter and there are a number of difficulties to be overcome — foremost among these is that the problem is numerically very large. A small discretization is necessary in order to resolve the features on the waveguide edges, and this not only increases the computation time in itself, but also for an FDTD algorithm the time-step must be reduced in order to keep within the Courant limit. This means that the calculation time becomes even larger. In addition, if the roughness is small in magnitude, then a large structure must be simulated in order to detect any attenuation at all.

In order to overcome these difficulties, we employed a wavelet-based FDTD algorithm [29], implemented on a parallel computer situated at the University of Karlsruhe. This cluster consists of 256 processors, each with a clock rate of 375 MHz and an available memory of 1 GB. In addition, we chose an unrealistically "rough" waveguide for our simulations. The enhanced roughness, together with the subdivision of the simulation region amongst the various processors, meant that the discretization could be chosen fine enough so that numerical algorithm was able to resolve the details of the sidewall, and that the simulation could be run long enough so that the attenuation could be confidently detected against the numerical noise.

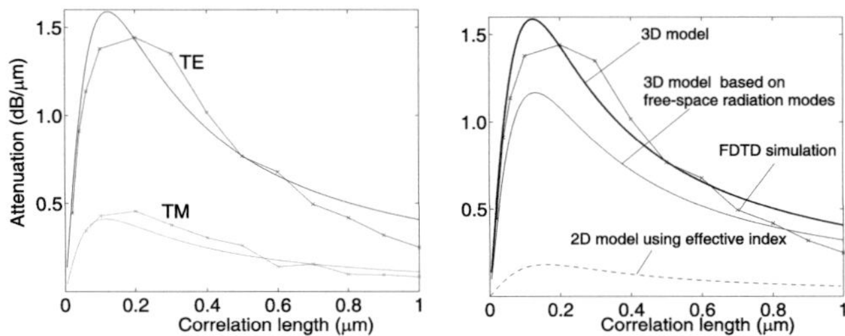

Fig. 3.6. Total attenuation as a function of correlation length D for a high-contrast square waveguide with width and height $2w = 2d = 0.365\,\mu$m and RMS sidewall perturbation of $\bar{\sigma} = 50$ nm, at frequency $f = 194$ THz. The waveguide has refractive index of $n_g = 3.03$ and lies on a glass substrate. a) Good agreement is seen between the 3D perturbation model (solid line) and the results of the FDTD simulation (crosses). Once again, the TE polarization exhibits a much larger attenuation than TM. b) Comparison of various methods used to calculate the attenuation of the TE mode. The approximation of the radiation modes by the free-space modes (such as shown in Figure 3.4b) compares favorably with these results, but results in an attenuation which is markedly smaller. The 2D effective index model predicts an attenuation which is an order of magnitude smaller than that predicted by other models.

Figure (3.6) shows the attenuation for a waveguide consisting of a square, high-contrast, waveguide with refractive index $n_g = 3.03$ lying on a substrate with refractive index 1.44. The waveguide width and height were $0.365\,\mu$m, and the RMS roughness was chosen to be $\bar{\sigma} = 50$ nm. The attenuation, calculated using Eq. (3.39), and directly using the FDTD algorithm, is shown plotted against the correlation length D. It can be seen that both the position of the maximum and the absolute value of the attenuation is predicted nicely by our model, for both TE and TM polarization. This is despite the relatively large value of $\bar{\sigma}$, which is assumed to be a small perturbation, and also despite the possible effect of the substrate, which is not taken into account in the perturbation model.

3.5.2 Comparison with Previous Coupled-Mode Models

It is important to compare the performance of the 3D coupled-mode model with the various models which have preceded it. This is shown, for the same waveguide as in the FDTD simulation, in Figure 3.6(b). Of particular interest is the comparison with the corresponding 2D model developed by Marcuse, for which an effective refractive index can be used in order to model a 3D waveguide. This is shown, for the same waveguide, in Figure 3.6(b). Although the position of the maximum is predicted nicely, the quantitative value of the attenuation is more than an order of magnitude lower than that predicted by the 3D model and by the FDTD algorithm. One reason for this is that the radiation in two dimensions is limited by one degree of freedom — the waveguide can only radiate in the plane perpendicular to the waveguide boundary. This model should continue to work well then for high aspect-ratio and slab waveguides, however the application to compact waveguides is limited.

It is also interesting to see how well our model will work if the radiation modes are approximated by their free-space equivalents, i.e. if the presence of the waveguide is ignored and the resulting free-space radiation modes (such as that shown in Figure 3.4) are used to determine the loss. It can be seen in Figure 3.6(b) that this procedure also gives reasonable results, although once again the absolute value is different to that predicted by the other 3D models. It is expected that this free-space approximation will work well for guides constructed out of low-contrast materials.

3.6 Measurement of Roughness and Attenuation

We conclude with an experimental investigation of the nature and effects of sidewall roughness in high-contrast waveguides. It has been seen that the attenuation depends on the statistics of the roughness autocorrelation function, which we have assumed up to this point to be modelled by an exponential function (3.37). In addition, in all the results presented so far it has been assumed that the waveguide sidewall perturbation is "chiselled" — i.e., that it does not change appreciably in the y-direction. The validity of these assumptions depends on the individual fabrication processes. To this end, we present experimental results on measurements of the surface topology of the roughness and the resulting attenuation of a typical high-contrast waveguide.

3.6.1 Fabrication of the Waveguide

Figure 3.7a) shows a SEM picture of a high-aspect ratio pedestal waveguide consisting of an InGaAsP guiding core, sandwiched between a InP pedestal and cap. The core used in the following measurements and calculations was 600 nm wide and 700 nm high, with pedestal height 1.5 μm and a cap of 300 nm thickness. The waveguide was written using e-beam lithography, then was transferred to a silicon nitride mask for the subsequent chemically assisted ion etching (CAIBE) process. Roughness occurs due to imperfections in the mask used to etch the structure, as well as to anisotropies in the etching process itself.

3.6.2 AFM Measurement of Surface Roughness

In order to measure the roughness the structure was cleaved along a vertical plane parallel to the waveguide. The tip of an atomic force microscope (AFM) then scanned the waveguide sidewall along the local y and z directions y' and z', and the deviation from the average sidewall position was measured over a local area of $4 \times 4\,\mu$m^2 (see Figure 3.7b). We used a commercial AFM (Autoprobe CP, Park Scientific Instruments, Sunnyvale, USA) with V-shaped silicon cantilevers with a nominal force constant of $0.35\,$Nm^{-1}. The tip was silicon nitride coated with a layer thickness of 10 nm. The images were taken in the

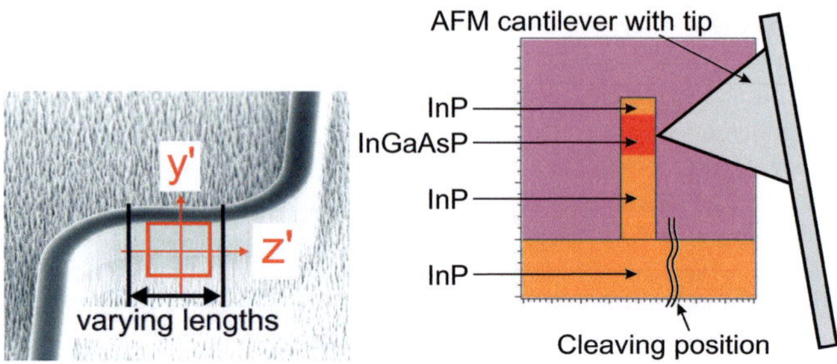

Fig. 3.7. a) SEM picture of a pedestal waveguide bend. b) Schematic of the pedestal waveguide and setup for measuring sidewall roughness.

contact mode of the AFM within the repulsive force regime and with a total normal force in the range of $(2...10) \times 10^{-8}$ N including capillary forces. The pyramidal structure of the tip would make measurement of a rib waveguide extremely difficult due to the contact of the tip with the cleaved edge.

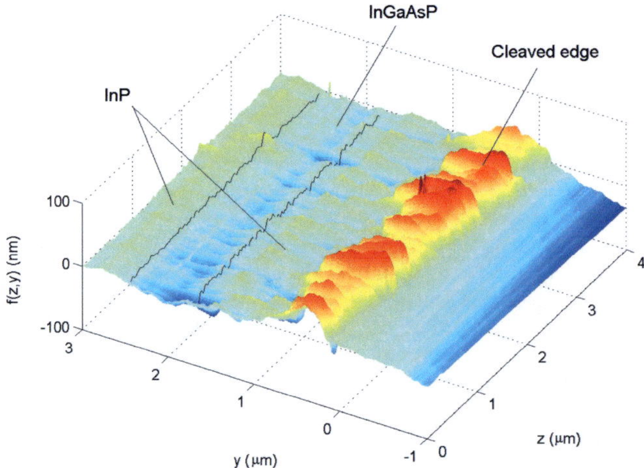

Fig. 3.8. AFM measurements of sidewall deformation function $f(z,y)$. The cleaved edge is visible as the large ridge centered around $y = 0.5\,\mu$m. The boundaries of the InGaAsP core are traced in black.

The measured sidewall deformation $f(z,y)$ of the waveguide is shown in Figure 3.8. The cleaved edge is clearly visible. The InGaAsP core of the waveguide can be seen as a trench-like depression in the middle of the measurement area — this indicates that for this fabrication process there is a certain undercut in InGaAsP. The measurement also confirms our earlier assumption that the sidewall roughness is essentially invariant in the y-direction.

The "chiselling" effect can most easily be seen by plotting the 2D autocorrelation function $R(u_z, u_y)$, as defined by

$$R(u_z, u_y) = \overline{f(z, y) f(z + u_z, y + u_y)} \tag{3.44}$$

where the bar indicates a two-dimensional integration over z and y, and so $R(u_z, u_y)$ can be interpreted as a single incidence of the ensemble of waveguide autocorrelations introduced in Section 3.4. From the measured data we isolate the points coming from the waveguide core, and normalize the data so that the mean value is zero and that the tilt arising from the unknown angle of the sample (with respect to the scanning plane of the AFM) is corrected. The resulting 2D autocorrelation function is shown in Figure 3.9(a).

It can be seen that the roughness is strongly correlated in the vertical direction, as is expected from a structure produced using an anisotropic etching process, and from these measurements the correlation length in the vertical direction is found to be greater than the height of the core. This result implies that a 1D correlation function of the general form (3.36) is sufficient for describing the surface roughness. By examining such a function $R(u)$ for each different y position on the waveguide wall (see Figure 3.9b), we find that the z correlation function fits reasonably well to the exponential model, with correlation length $D = (56 \pm 14)$ nm and an RMS deviation of $\bar{\sigma} = (5 \pm 1)$ nm. Interestingly, the roughness is also periodically correlated in the z direction, and in fact fits well to the model

$$R(u_z) = \bar{\sigma}_1^2 \exp(-|u_z|/D_1) + \bar{\sigma}_2^2 \cos(2\pi u_z/D_2) , \tag{3.45}$$

where the exponential correlation length is $D_1 = (54 \pm 21)\,\text{nm}$ with an RMS deviation of $\bar{\sigma}_1 = (4.9 \pm 1.5)\,\text{nm}$, and the length of the periodic variation is $D_2 = 610\,\text{nm}$, with an RMS amplitude of approximately $\bar{\sigma}_2 = 1.0\,\text{nm}$. Although it is possible that such periodic variation will lead to Bragg reflection at certain frequencies, it seems likely that any such effect will be swamped by the much stronger exponential part of the autocorrelation.

3.6.3 Measured Waveguide Loss and Comparison with Model

In our experiment, the waveguide attenuation was measured using a set of dogleg-shaped waveguides, each consisting of two bends separated by a straight segment of all different lengths L_{str} (see Figure 3.7a). The Hakki-Paoli method [45] was then used to calculate the round-trip attenuation from the contrast of Fabry-Perot resonances arising from the reflection at the cleaved facets of the waveguides. Bend losses and the reflection factors were unknown, but could be eliminated using samples of different lengths L_{str}. The loss per unit length can then be deduced by plotting the round-trip amplitude transmission factor a_{RT} against L_{str}. From the slope of this curve a power attenuation of $\alpha_\mu = (7.0 \pm 1)\,\text{dB / mm}$ could be deduced for the lowest order TE mode, and $2\alpha_\mu = (4.0 \pm 1)\,\text{dB / mm}$ for the lowest-order TM mode.

Although the semi-analytical model discussed in the previous sections is designed for a different structure to the experimental guide, it may be expected that good results can be

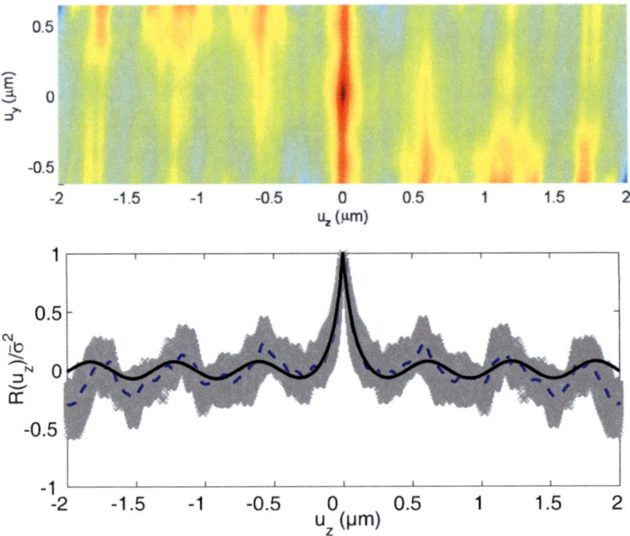

Fig. 3.9. a) Two-dimensional autocorrelation function $R(u_z, u_y)$ of the measured sidewall roughness data for a pedestal waveguide. The "chiselling" effect (invariance in the y-direction) is clearly visible. b) Calculated one-dimensional autocorrelation functions $R(u_z)$ for the complete set of different vertical positions over the wall (shaded grey). The mean value of the set is shown as a dashed line, the fit to the model (3.37) is shown as a solid line. The periodic variation is clearly visible, with period 610 nm.

obtained by treating the radiation modes of the pedestal waveguide as being approximated by those of a rectangular waveguide with the same dimensions and refractive index as the pedestal core. Using the measured values for the correlation length D_1 and RMS variation $\bar{\sigma}_1$, the attenuation was predicted using the semi-analytical model. The results are shown in Figure 3.11. It was found that the predicted attenuation for the pedestal waveguide using the 3D semi-analytical model was $(5.7 \pm 4.4)\,\mathrm{dB\,/\,mm}$ for the TE polarization and $(1.5 \pm 1.2)\,\mathrm{dB\,/\,mm}$ for the TM polarization. The large error estimates arise due to the uncertainty in the measurement of the correlation function: Even a small variation in either correlation length or RMS variation can cause a dramatic increase or decrease in attenuation. In order to achieve an uncertainty of less than 10% of the attenuation, the RMS roughness must be known to an accuracy of better than 2 Angstrom. Additional uncertainty comes from the unknown waveguide material loss which we assumed to be negligible, as well as from roughness-induced leakage into the substrate, which is not taken into account by the semi-analytical model. In any case, it can be seen that these values predicted for the attenuation come very close to predicting the attenuation. By contrast, if a 2D effective index method is used, then the predicted values for the loss are 0.53 ± 0.45

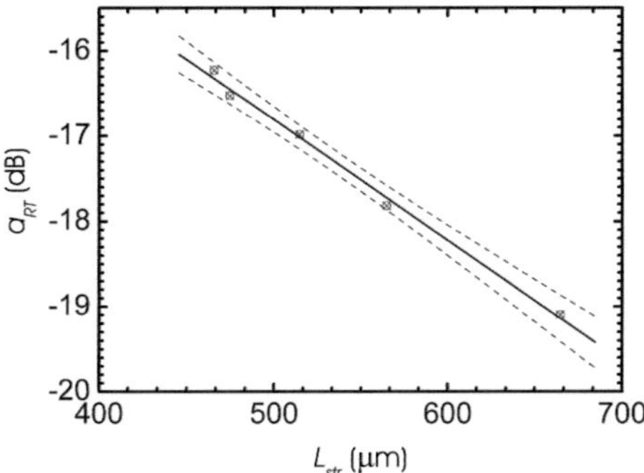

Fig. 3.10. Measured round-trip amplitude transmission factor a_{RT} ($\otimes$) for the lowest-order TE mode of a pedestal waveguide, as a function of waveguide length L_{str}. (——) linear fit to measurements, (– – –) 95 % confidence limits for the fit

dB/mm for TE polarization, and 0.20 ± 0.17 dB/mm for TM polarization, and so are an order of magnitude too small.

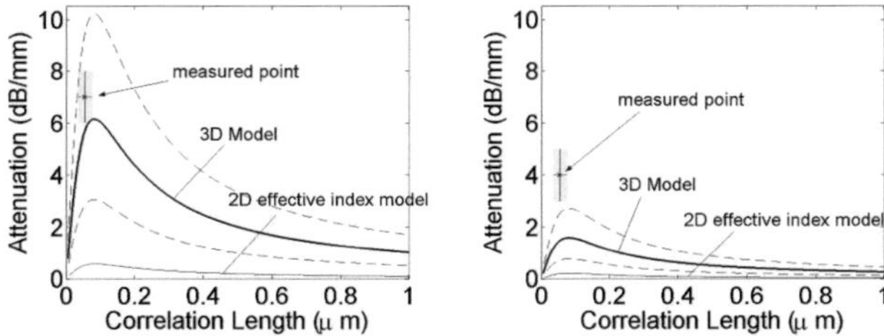

Fig. 3.11. Attenuation as a function of correlation length D for the InGaAsP/InP pedestal waveguide, for (a) TE and (b) TM polarizations. The dashed lines show the error limits due to the uncertainty in the measurement of the roughness characteristics. It can be seen that in both cases the measured attenuation is predicted reasonably well by the 3D semi-analytical model. The 2D effective index model predicts an attenuation which is an order of magnitude smaller.

3.7 Summary

We have introduced a semi-analytical model for the prediction of attenuation due to sidewall deformations in high-contrast dielectric waveguides. We have seen that by taking into account three-dimensional radiation modes a much better estimate of roughness losses can be achieved than for two-dimensional models. We note that the radiation modes themselves are an important development and could be employed in many other applications in integrated optics. The formalism also gives physical insight for observed phenomena — namely, the difference in attenuation between TE and TM guided mode polarizations, and the fact that the dominant mechanism for attenuation due to sidewall roughness is radiation rather than reflection. The analytical treatment of the problem allows us to derive guidelines for designing the cross section of low-loss waveguides. We have also investigated the characteristics of sidewall roughness using atomic force microscopy on a fabricated waveguide, and have found that the exponential model of sidewall roughness fits well, despite the presence of periodic correlations which can arise in the fabrication process used here. The parameters gained from these measurements can be used for accurate prediction of roughness-induced losses in high-contrast integrated optical waveguides.

Chapter 4

Minimizing Waveguide Loss by Design: Ideal Contour Trajectories

4.1 Introduction

Planar lightwave circuits (PLC) are commonly fabricated by locally changing the refractive index on the surface of a substrate material, or by removing material from a layered substrate. In both cases, the refractive index profile in the vertical direction can be approximated by an effective index method [19], and the design problem then reduces from a three-dimensional one to two dimensions. Integrated optical devices are then represented by a device domain D_{dev} in the substrate plane, in which the effective index is increased compared to the surrounding ambient domain D_{amb}. From a mathematical point of view, the design of the integrated optical device can be considered as a quest for an appropriate contour trajectory C_{cont} that separates the device domain from the ambient domain. Light is confined to the high-index device domain D_{dev} by reflection at the domain boundary, whereby losses are unavoidable if this boundary is curved.

The fabrication of semiconductor-based high index-contrast PLC comprises two main steps: First, a mask defining the device domain and the ambient domain is structured on top of a layered substrate material. This is usually done by means of lithographic tools such as electron-beam lithography or optical lithography. Second, material is removed from the domain that is not protected by the mask, usually by using dry etching techniques. Both of these fabrication steps have physical limitations that can severely impair the performance of high index-contrast devices: The resolution of lithographic processes is generally limited by proximity effects and/or the resist chemistry. As a consequence, the contour trajectory of the real structure can considerably deviate from the designed trajectory. Since high index-contrast devices are usually an order of magnitude smaller than their low index-contrast counterparts, limited resolution is an important constraint. Furthermore, the sidewalls of etched semiconductor structures usually exhibit a certain surface roughness which leads to scattering loss in addition to the unavoidable curvature induced loss.

Understanding these physical limitations of the fabrication process, two important design goals can be formulated for the device contour: First, radiation loss originating from curved and/or rough sidewalls should be reduced to a minimum. Second, the contour trajectory should be robust with respect to a fabrication process with limited resolution. In this chapter, we will derive a class of contour trajectories that fulfill both of these requirements in an optimal way. For demonstration, we design compact, low-loss waveguide bends, and we compare these bends to conventional designs. The results are verified by simulations and by experiments.

Parts of the following chapter have been published in a journal article [J4], see pp. 199.

4.2   Contour Trajectories for Minimum Radiation Loss

Theoretical investigations of scattering in HIC waveguides have shown that roughness-induced radiation loss is mainly dictated by the magnitude of the field strength at the rough surface, see Eqs. (3.40) and (3.41). As described in Section 3.4.2, scattering loss of straight waveguides can be lowered by increasing the waveguide width [24] – possibly at the expense of making the straight waveguide multimoded. It is then of high importance to design all components for minimum coupling between the fundamental and higher-order modes.

For bent sections, whispering gallery propagation of light along the outer contour reduces interaction with rough sidewalls to a minimum and allows for low loss. Unavoidable curvature-induced radiation loss may then be further minimized by an appropriate contour trajectory. By optimizing the shape of the waveguide bend, it is possible to operate low-loss overmoded waveguides in a single-mode way. This shall be demonstrated in the following. We first derive a class of ideal outer contour trajectories analytically. In a second step, numerical optimization is used to obtain single-mode operation of waveguide bends that are connected to overmoded feed waveguides.

4.2.1   Class of Ideal Bend Contour Trajectories

The contours of the waveguide bend should smoothly connect the ends of two straight multimode waveguides of widths w_0 and w_1, see Fig. 4.1(a). Thus, the inner and the outer contour are each fixed at two points in the (x, z)-plane, where the contour angles are φ_0, φ_1.

Outer Contour

For the outer contour, we assume that the angle φ between the z-axis and the local tangent is monotonic with the arc length s and can thus be chosen as a curve parameter, see Fig. 4.1(a). The angle φ ranges from φ_0 in the starting point $\mathbf{r}_{\mathrm{oc}}(\varphi_0) = (x_0, z_0)$ to φ_1 in the ending point $\mathbf{r}_{\mathrm{oc}}(\varphi_1) = (x_1, z_1)$, thus ensuring kink-free transitions between the bent and the straight sections. The function $s(\varphi)$ denotes the arc length, and $\kappa(\varphi) = \mathrm{d}\,\varphi/\mathrm{d}\,s$ is the local curvature of the contour. The outer contour can then be represented as

$$\mathbf{r}_{\mathrm{oc}}(\varphi) = \mathbf{r}_{\mathrm{oc}}(\varphi_0) + \int_{\varphi_0}^{\varphi} [\kappa(\widetilde{\varphi})]^{-1} \begin{pmatrix} \sin(\widetilde{\varphi}) \\ \cos(\widetilde{\varphi}) \end{pmatrix} \mathrm{d}\widetilde{\varphi} \qquad (4.1)$$

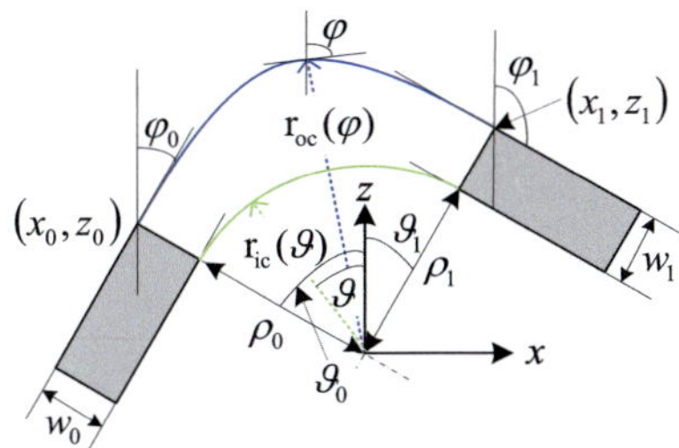 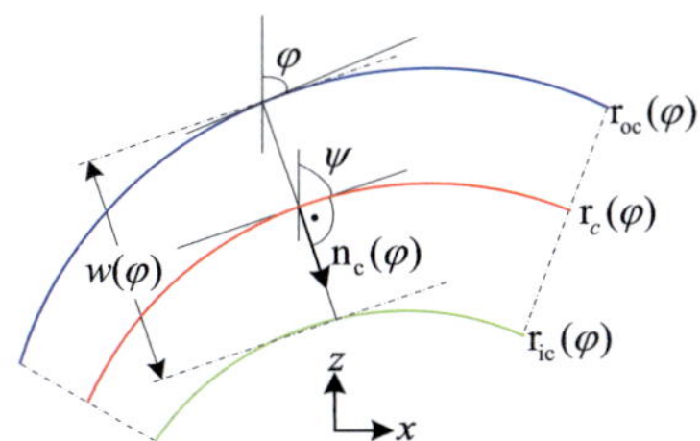

(a) The outer contour $\mathbf{r}_{\mathrm{oc}}(\varphi)$ (blue) links the points (x_0, z_0) and (x_1, z_1) in the (x, z)-plane. The associated angles with respect to the z-axis are φ_0 (φ_1) in the start (end) point. Similar conditions hold for the inner contour $\mathbf{r}_{\mathrm{ic}}(\varphi)$ (green).

(b) The curve can be defined by the outer contour $\mathbf{r}_{\mathrm{oc}}(\varphi)$ (blue) and the width $w(\varphi)$ which is measured perpendicularly to the center line $\mathbf{r}_{\mathrm{c}}(\varphi)$ of the waveguide (red). The local direction of the center line is given by the angle ψ with respect to the z-direction.

Fig. 4.1. Waveguide Bends

To minimize the radiation loss for whispering gallery mode propagation, the power loss coefficient α as a function of the local contour curvature κ has to be known. Taking into account both curvature-induced radiation and scattering due to sidewall roughness, the exact relationship between α and κ is difficult to predict theoretically. Therefore, we assume a local approximation of the form $\alpha \propto |\kappa|^q$ ($q > 1$). From experiments [25], $2 \leq q \leq 3$ can be estimated for SOI strip waveguides, and for the remainder of this chapter, we use $q = 2$. Minimizing the radiation losses means finding an outer contour $r_{\mathrm{oc}}(\varphi)$ such that the integral

$$I = \int_{\varphi_0}^{\varphi_1} |\kappa(\varphi)|^q \, \frac{\mathrm{d}s}{\mathrm{d}\varphi} \, \mathrm{d}\varphi \tag{4.2}$$

is minimal. Two additional constraints have to be introduced to guarantee that the curve links the given starting and ending points,

$$\begin{pmatrix} x_1 \\ z_1 \end{pmatrix} - \begin{pmatrix} x_0 \\ z_0 \end{pmatrix} = \int_{\varphi_0}^{\varphi_1} [\kappa(\widetilde{\varphi})]^{-1} \begin{pmatrix} \sin(\widetilde{\varphi}) \\ \cos(\widetilde{\varphi}) \end{pmatrix} \mathrm{d}\widetilde{\varphi}, \tag{4.3}$$

Now the variational problem can be formulated in terms of the arc length s, the local angle $\varphi(s)$ as a function of arc length, and its first derivative $\varphi'(s) = \mathrm{d}\varphi/\mathrm{d}s$. The corresponding Lagrangian reads

$$L\left(s, \varphi, \varphi', \lambda_1, \lambda_2\right) = |\varphi'|^q + \lambda_1 \left(\sin\varphi - \Delta x\right) + \lambda_2 \left(\cos\varphi - \Delta z\right), \tag{4.4}$$

where $\Delta x = x_1 - x_0$ and $\Delta z = z_1 - z_0$. Inserting Eq. 4.4 into Euler's differential equation,

$$\frac{\partial L}{\partial \varphi} - \frac{\mathrm{d}}{\mathrm{d}s}\left(\frac{\partial L}{\partial \varphi'}\right) = 0, \tag{4.5}$$

we obtain the curvature κ as a function of φ,

$$\kappa\left(\varphi\right) = C_1 \left[\cos\left(\varphi - C_2\right) + C_3\right]^{1/q}, \tag{4.6}$$

where C_1, C_2, and C_3 are constants that have to be determined from the boundary conditions. Using Eq. 4.1, the general solution for the outer contour can be written as

$$\mathbf{r}_{\mathrm{oc}}(\varphi) = \mathbf{r}_{\mathrm{oc}}(\varphi_0) + C_1^{-1} \int_{\varphi_0}^{\varphi} \frac{1}{\left[\cos\left(\widetilde{\varphi} - C_2\right) + C_3\right]^{1/q}} \begin{pmatrix} \sin(\widetilde{\varphi}) \\ \cos(\widetilde{\varphi}) \end{pmatrix} d\widetilde{\varphi} \tag{4.7}$$

Two of the three constants C_1, C_2, C_3 are needed to fix the ending point. The remaining degree of freedom can be used to adjust the ratio of the outer contour's curvatures in these points and is subject to numerical optimization.

The contour curve, Eq. (4.7), leads to a minimum (and not just to an extremum) of the Integral I in Eq. (4.2). This can be shown mathematically be exploiting the fact that for $\kappa \neq 0$ and $q > 1$, the integrand $|\kappa|^q \frac{ds}{d\varphi} = |\kappa|^{q-1}$ of Eq. (4.2) is a convex function of κ.

Inner Contour

The shape of the inner contour has to be optimized numerically such that the fundamental mode of the straight waveguide is transformed into a whispering gallery mode with minimal losses. To define the inner boundary in the following, we will use a generic taper function $g(u)$ that is defined by

$$g(u) = \sum_{\nu=1}^{\nu_{\max}} a_\nu \cos(2\pi\nu u), \tag{4.8}$$

where $a_\nu = 1$ and $a_2 \ldots a_{\nu_{\max}}$ ($\nu_{\max} \leq 5$ typically) are free parameters and allow to vary the shape of the inner bend contour. Note that for $u = 0$ and $u = 1$, all derivatives $d^\mu g(u)/du^\mu$ of odd order μ vanish.

For small radii of curvature, we use a direct parametrization of the inner contour. In the following, the origin of the coordinate system is assumed to be at the crossing point of the lines that define the ends of the straight sections, see Fig. 4.1(a). Then a convenient parametrization of the inner contour based on the angle ϑ between the local position vector and the z-axis can be used,

$$\mathbf{r}_{\mathrm{ic}}(\vartheta) = \rho(\vartheta) \begin{pmatrix} \sin\left(\vartheta\right) \\ \cos\left(\vartheta\right) \end{pmatrix}, \tag{4.9}$$

where the curve parameter ϑ ranges from ϑ_0 to ϑ_1. The angle-dependent distance $\rho(\vartheta)$ from the origin can be defined by

$$\rho(\vartheta) = \rho_0 + (\rho_1 - \rho_0)\frac{g\left(\frac{\vartheta - (\vartheta_1 + \vartheta_0)/2}{\vartheta_1 - \vartheta_0}\right) - g\left(-0.5\right)}{g\left(0.5\right) - g\left(-0.5\right)}, \tag{4.10}$$

where ρ_0 (ρ_1) is the distance of the starting (ending) point from the origin. Note that this function has zero slope in the starting and the ending point, ensuring kink-free transitions at both ends of the inner bend contour.

For large radii of curvature, rather than assuming an independent inner bend contour, it can be useful to optimize the width $w(\varphi)$ of the bent waveguide as a function of the local outer contour direction φ. In this case, the inner contour $\mathbf{r}_{\mathrm{ic}}(\varphi)$ has to be reconstructed from a given outer contour $\mathbf{r}_{\mathrm{oc}}(\varphi)$ and a given width taper function $w(\varphi)$. A waveguide width taper of the form

$$w(\varphi) = w_0 + (w_1 - w_0)\frac{g\left(\frac{\varphi-(\varphi_1+\varphi_0)/2}{\varphi_1-\varphi_0}\right) - g\left(-0.5\right)}{g\left(0.5\right) - g\left(-0.5\right)} \tag{4.11}$$

has proven useful. The function $w(\varphi)$ has zero slope in the starting and the ending point, ensuring kink-free transition also for the inner bend contour.

To reconstruct the inner contour, let us first introduce the centerline $\mathbf{r}_{\mathrm{c}}(\varphi) = \left(\mathbf{r}_{\mathrm{oc}}(\varphi) + \mathbf{r}_{\mathrm{ic}}(\varphi)\right)/2$ of the waveguide. The width is measured perpendicularly to $\mathbf{r}_{\mathrm{c}}(\varphi)$, and the outer and inner contours can be be written as

$$\mathbf{r}_{\mathrm{oc}}(\varphi) = \mathbf{r}_{\mathrm{c}}(\varphi) - \frac{w(\varphi)}{2}\,\boldsymbol{n}_{\mathrm{c}}(\varphi), \tag{4.12}$$

$$\mathbf{r}_{\mathrm{ic}}(\varphi) = \mathbf{r}_{\mathrm{c}}(\varphi) + \frac{w(\varphi)}{2}\,\boldsymbol{n}_{\mathrm{c}}(\varphi), \tag{4.13}$$

where $\mathbf{n}_{\mathrm{c}}(\varphi)$ is a unit vector that is normal to the centerline and points towards the center of the curve, see Fig. 4.1(b). The local direction of the centerline is described by the angle ψ with respect to the z-direction. To reconstruct the inner contour from a given outer contour $\mathbf{r}_{\mathrm{oc}}(\varphi)$ and a given width taper $w(\varphi)$, the centerline direction $\psi(\varphi)$ is first calculated from

$$\sin\left(\varphi - \psi\right) = \frac{w'(\varphi)\kappa\left(\varphi\right)}{2}, \tag{4.14}$$

where $w'(\varphi) = \mathrm{d}\,w/\,\mathrm{d}\,\varphi$. The vector $\mathbf{n}_{\mathrm{c}}(\varphi)$ can the be written as

$$\mathbf{n}_{\mathrm{c}}(\varphi) = \begin{pmatrix} \cos\psi(\varphi) \\ -\sin\psi(\varphi) \end{pmatrix}, \tag{4.15}$$

and the inner contour is then obtained from

$$\mathbf{r}_{\mathrm{ic}}(\varphi) = \mathbf{r}_{\mathrm{oc}}(\varphi) + w(\varphi)\mathbf{n}_{\mathrm{c}}(\varphi). \tag{4.16}$$

Numerical Optimization

The inner and the outer contour curves finally have to be numerically optimized for lossless transitions between the fundamental mode of the straight waveguides and the whispering gallery mode of the curved section. The exact shapes of the contour curves depend on a set of shape parameters. A Nelder-Mead optimization based on a two-dimensional (2D) FDTD algorithm is used to find optimum shape parameters.

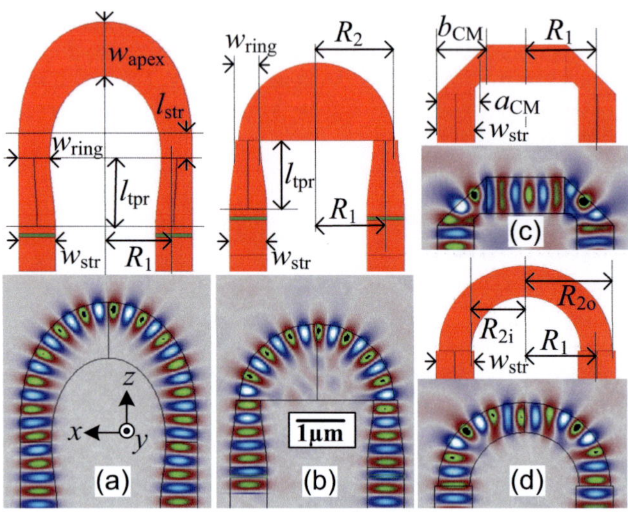

Fig. 4.2. Different bend layouts for $R_1 = 1.5\,\mu$m and $w_{\mathrm{str}} = 0.8\,\mu$m and the corresponding field plots (E_y-component of the 2D FDTD simulation). (a) Contour optimized bend (COB): $l_{\mathrm{tpr}} = 1.5\,\mu$m, $w_{\mathrm{apex}} = 1.2\,\mu$m, $w_{\mathrm{ring}} = 0.677\,\mu$m, $l_{\mathrm{str}} = 0.048\,\mu$m. (b) Semicircular disc with taper, $l_{\mathrm{tpr}} = 1.5\,\mu$m, $w_{\mathrm{ring}} = 0.524\,\mu$m, $R_2 = 1.695\,\mu$m. (c) Corner mirrors: $a_{\mathrm{CM}} = 0.92\,\mu$m, $b_{\mathrm{CM}} = 1.06\,\mu$m. (d) Offset circular bend: $R_{2\mathrm{i}} = 1.16\,\mu$m, $R_{2\mathrm{o}} = 1.84\,\mu$m.

4.2.2 Design of SOI Waveguide Bends

We developed contour-optimized 180°-bends between two straight overmoded SOI strip waveguides which have a fixed width of $w_{\mathrm{str}} = 0.8\,\mu$m, see[1] Fig. 4.2 (a). Such bends can be used for waveguide-based delay-lines [118]. The waveguides consist of Si ribs (refractive index $n_{\mathrm{Si}} = 3.48$, height $h = 340\,$nm) optically isolated from the Si substrate by an underlying $3\,\mu$m thick SiO_2 layer ($n_{SiO_2} = 1.44$), and are surrounded by air. At a wavelength of $\lambda = 1.55\,\mu$m, each of the straight waveguides supports three guided TM modes (dominant E_y-component) and three guided TE modes. The fundamental TM mode was launched in the left waveguide, and the power coupled into the fundamental TM mode of the right waveguide was maximized using 2D FDTD modelling. The discretization was $\Delta x = \Delta z = 0.02\,\mu$m, and the normalized time step was $c\Delta t = 0.01\,\mu$m, where c is the velocity of light. The contour-optimized bend (COB) comprises a straight tapered waveguide section of fixed length l_{tpr} and the curved section. The curved outer contour is defined according to Eq. (4.7) and is preceded by a straight section of length l_{str}. The inner contour is defined according to Eqs. (4.8) and (4.10), with $\nu_{\mathrm{max}} = 2$ shape parameters. The transmission is optimized by varying the shape parameters a_1 and a_2, the

---

[1]Please note that the x-axes in Fig. 4.1(a) point to the right, whereas in Fig. 4.2 the positive x-direction points to the left. This change is necessary to ensure that (x, y, z) represents a right-handed coordinate system in Fig. 4.2.

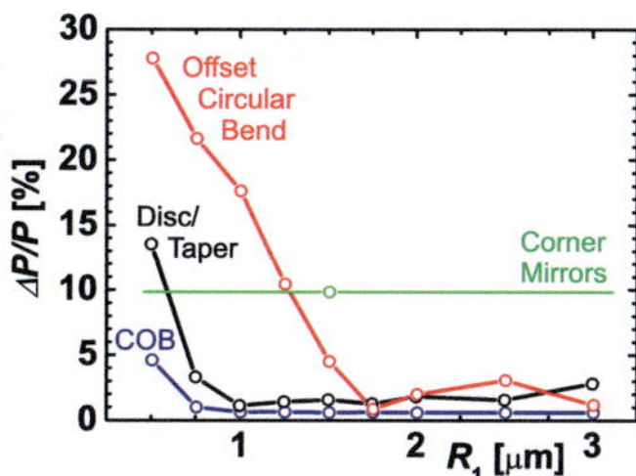

Fig. 4.3. Normalized power loss for the contour optimized bend (COB), the disc/taper configuration, the laterally offset circular bend, and the pair of corner mirrors.

initial curvature of the outer contour, and the parameters l_{str} and w_{ring}. The waveguide width w_{apex} at the curve's apex point is fixed. The field plot in Fig. 4.2 (a) shows nearly perfect transition between the fundamental feed waveguide mode and the fundamental whispering gallery mode. No higher-order modes are excited.

The COB are compared to other bend shapes whose geometries and 2D FDTD field plots are shown in Fig. 4.2. In Fig. 4.2 (b), the bend consists of a semicircular disc of radius R_2. To match the fundamental waveguide mode and the whispering gallery mode, a taper of length l_{tpr} and final width of w_{ring} has been introduced. The parameters R_2 and w_{ring} are varied to maximize the transmission. For the double corner mirror configuration shown in Fig. 4.2 (c), the parameters a_{CM} and b_{CM} are varied, and the offset circular bend in Fig. 4.2 (d) is optimized by varying $R_{2\text{i}}$ and $R_{2\text{o}}$. Apart from weak radiation in the center of the disc, the field plot for the disc/taper configuration shows very good light guidance. The offset circular bend shows moderate radiation loss, and the corner mirrors suffer from strong radiation and substantial mode conversion. For $R_1 = 1.5\,\mu\text{m}$, we obtain loss values of 0.6%, 1.6%, 4.5% and 9.8% for the COB, the disc/taper configuration, the offset circular bend and the pair of corner mirrors, respectively.

Low insertion loss is retained for the COB when reducing the distances between the straight waveguide sections: Fig. 4.3 depicts the minimized relative power losses between the fundamental TM modes of the straight waveguide as a function of R_1 for the four different bend types. For the COB, the power loss remains below 1% for $R_1 > 0.75\,\mu\text{m}$, and only increases for unrealistically low values of $R_1 = 0.5\,\mu\text{m}$. Our curve design thus permits mean radii of curvature R_1 that are smaller than the waveguide width $w_{\text{str}} = 0.8\,\mu\text{m}$ without substantial mode conversion. The disc/taper configuration has larger losses for $R_1 < 1\,\mu\text{m}$. For the offset circular bends, losses increase considerably for decreasing R_1. The minimum for $R_1 = 1.75\,\mu\text{m}$ is due to a specific interference of different modes of the bent waveguide at the transition to the straight output section which occurs only for this special wavelength. For the corner mirrors, reducing R_1 can be achieved by making the straight waveguide section between the corner mirrors shorter. The bend loss is therefore considered to be independent of R_1, as is indicated in Fig. 4.3 by the horizontal line passing through the data point for $R_1 = 1.5\,\mu\text{m}$.

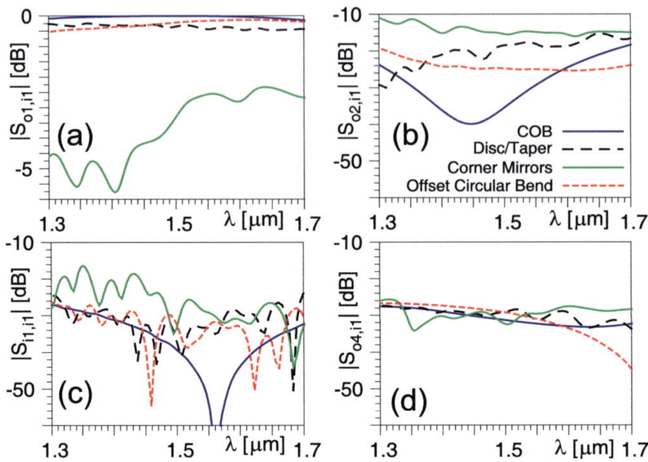

Fig. 4.4. S-parameters obtained from a 3D FDTD simulation. (a) Transmission; (b) Mode conversion; (c) Reflection; (d) Polarization conversion.

To validate the results of the 2D COB design, a broadband three-dimensional (3D) FDTD simulation of the optimized structure is performed. The discretization was $\Delta x = \Delta y = \Delta z = 0.025\,\mu$m, and the normalized time step was chosen to be $c\Delta t = 0.012\,\mu$m. The resulting scattering matrix parameters connecting the various input and output modes for the case of a moderate $R_1 = 1.5\,\mu$m are shown in Fig. 4.4. The fundamental TM mode at the input (output) port is labelled with a subscript i1 (o1), the first higher-order TM mode is denoted by i2 (o2), and the fundamental TE mode is subscripted with i4 (o4). For the COB, the transmission $S_{o1,i1}$ is better than $-0.2\,$dB ($-0.1\,$dB) over a bandwidth of more than 300 nm (100 nm), and coupling to the first higher-order TM mode $S_{o2,i1}$ is below $-20\,$dB ($-30\,$dB). Both, reflection $S_{i1,i1}$ and polarization conversion $S_{o4,i1}$, are below $-30\,$dB over more than 300 nm. For the disc/taper configuration, the transmission $S_{o1,i1}$ is around $-0.35\,$dB, and for the offset circular bend, $S_{o1,i1}$ ranges from $-0.6\,$dB to $-0.2\,$dB for wavelengths λ between $1.30\,\mu$m and $1.70\,\mu$m. In terms of coupling to the first higher-order TM mode ($S_{o2,i1}$), the COB is superior for $1.32\,\mu$m $< \lambda < 1.60\,\mu$m, and concerning reflection at the input port ($S_{i1,i1}$), the COB performs best around the design wavelength $\lambda = 1.55\,\mu$m. Polarization conversion ($S_{o4,i1}$) does not vary much among the four different structures. We note that both for the COB and for the offset circular bends, the results from the 2D and from the 3D FDTD simulations are in fair agreement. For the disc/taper configuration and the corner mirrors, however, the transmission $S_{o1,i1}$ predicted by the 3D FDTD simulation is much worse than the one predicted by the 2D FDTD model. This is caused by an increase in radiation loss which might be due to an additional degree of freedom in the manifold of radiation modes for the 3D model [86]. For the COB and the offset circular bends, radiation losses only play a minor role.

4.2.3 Fabrication and Measurement

Waveguides with COB structures were fabricated on SOITEC Smart-Cut^{TM} SOI wafers. A 70 nm thick Si_3N_4 layer was deposited by plasma-enhanced chemical vapour deposition and patterned by direct electron beam writing and dry etching. The structures were transferred to the SOI device layer using reactive ion etching (RIE). Thinning and cleaving was done with standard techniques. TM-polarized light was coupled to the facets using polarization-maintaining lensed fibres. An SEM micrograph of a fabricated bend is shown in the inset of Fig. 4.5 (right). We fabricated meander-shaped waveguides with 2, 4, 6 and 40 bends, and measured the transmission spectra, Fig. 4.5 (left). In Fig. 4.5 (right) the transmission vs. the number of bends is depicted. A straight line was fitted to the data. From the slope we deduce an insertion loss of 0.35 dB for a 180° COB. The loss in excess of the FDTD results is attributed to roughness and to fabrication inaccuracies. Still, this is an extremely low loss regarding the fact that the etch process was not optimized for sub-μm structures. These losses are comparable with results from highest performance CMOS-technology single-mode bends [25, 111, 118].

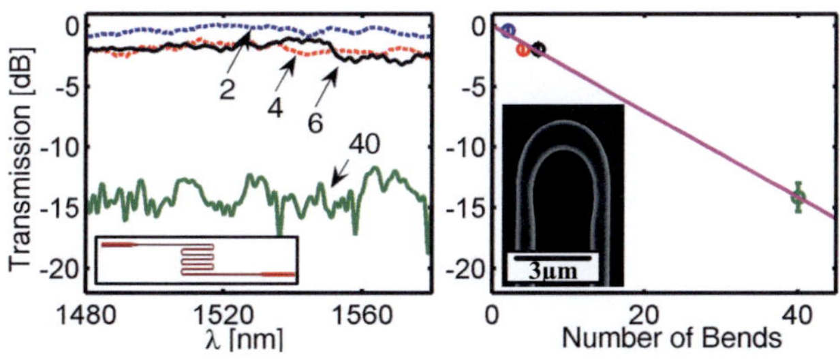

Fig. 4.5. Left: Measured transmission spectra for different meander waveguides with 2, 4, 6, and 40 bends. Right: Transmission losses vs. number of bends together with a linear fit. Inset left: Meander waveguide with 6 bends. Inset right: SEM micrograph of a fabricated bend.

4.3 Robust Contour Trajectories for Limited Resolution

We will now investigate the sensitivity of contour trajectories with respect to the resolution of the fabrication process. We first introduce a phenomenological model for the fabrication process, based on which we derive a class of ideal contour trajectories. In the following, $\mathbf{r}(t) = (x(t), z(t))^T$ denotes an arbitrary representation of the designed contour trajectory $C_{\text{cont}}^{(\text{des})}$ with respect to a curve parameter t. $C_{\text{cont}}^{(\text{des})}$ is the designed boundary between the device domain D_{dev} and the ambient domain D_{amb}, see Fig. 4.6. $C_{\text{cont}}^{(\text{real})}$ is the real contour that deviates by δ from $C_{\text{cont}}^{(\text{des})}$ due to resolution limitations. The angle φ defines the local direction of the contour, and (u, v) are the coordinates with respect to a local coordinate system with origin $\mathbf{r}(t_0)$ and the u-axis oriented tangentially to the curve.

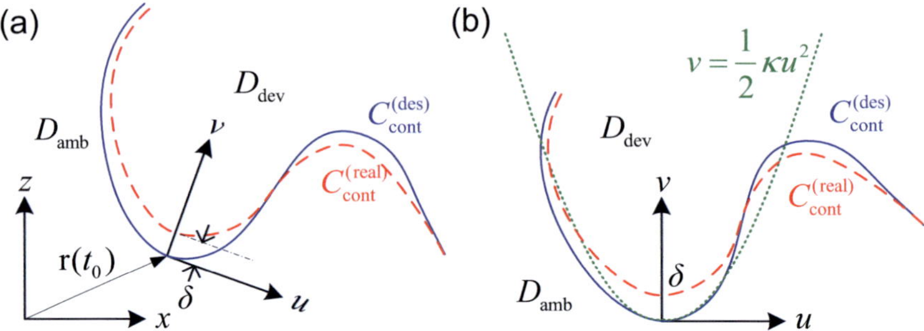

Fig. 4.6. (a) Designed contour curve $C_{\text{cont}}^{(\text{des})}$ (solid) separating the designed device domain D_{dev} from the designed ambient domain D_{amb}. For a fabrication process with limited resolution, the real contour curved $C_{\text{cont}}^{(\text{real})}$ (dashed) deviates from the designed trajectory $C_{\text{cont}}^{(\text{des})}$. The local deviation is named δ. The angle φ defines the local direction of the contour curve, and (u, v) are coordinates with respect to a local coordinate system. (b) The designed contour curve $C_{\text{cont}}^{(\text{des})}$ can be locally approximated to second order by a parabola $v = \frac{1}{2}\kappa u^2$ in the (u, v) coordinate system, where κ denotes the curvature of $C_{\text{cont}}^{(\text{des})}$ in the point $\mathbf{r}(t_0)$.

4.3.1   Fabrication Process Model

Any lithographic process exhibits proximity effects, which lead to a blurring of the exposed structure with respect to the designed one. For optical lithography, the resolution is limited by diffraction, and the minimum resolvable feature size corresponds roughly to the diameter d_{A} of the Airy disc, $d_{\text{A}} = 1.22\lambda/\text{NA}$, where λ is the wavelength of coherent illumination, and NA denotes the numerical aperture of the system. For direct-write electron beam lithography, proximity effects arise mainly from scattering of primary electrons in the resist and from backscattering of secondary electrons from the substrate material. Let $g^{(\text{des})}(x, z)$ represent the normalized designed exposure profile that would be obtained by an ideal lithographic system of infinite resolving power,

$$g^{(\text{des})}(x, z) = \begin{cases} 1 & \text{for } (x, z) \in D_{\text{exp}} \\ 0 & \text{else} \end{cases}, \tag{4.17}$$

where the exposed domain D_{exp} depends on the lithography process. For a negative-tone process, the device domain is exposed, $D_{\text{exp}} = D_{\text{dev}}$, and for a positive-tone process, the ambient domain is exposed, $D_{\text{exp}} = D_{\text{amb}}$. Due to proximity effects, the real exposure profile $g^{(\text{real})}(x, z)$ is blurred with respect to $g^{(\text{des})}(x, z)$, which can be phenomenologically described by convolving the designed layout with a two-dimensional imaging kernel $h(x, z)$,

$$g^{(\text{real})}(x, z) = \iint\limits_{-\infty}^{+\infty} g^{(\text{des})}(x - \xi, z - \eta) h(\xi, \eta) \, \mathrm{d}\xi \, \mathrm{d}\eta. \tag{4.18}$$

The imaging kernel shall be assumed to be normalized,

$$\iint\limits_{-\infty}^{+\infty} h(x, z)\, \mathrm{d}x\, \mathrm{d}z = 1. \tag{4.19}$$

The resist can be modelled as a limiter: A negative-tone (positive-tone) resist will be removed at all points (x, z) where the exposure dose falls below (exceeds) a certain threshold g_{th}. The threshold depends on the development process. The real contour trajectory comprises the set of points for which the exposure dose is equal to the development threshold, and is defined by the implicit equation

$$g^{(\text{real})}(x, z) - g_{\text{th}} = 0. \tag{4.20}$$

It should be mentioned that the etching process may erode the etching mask and thereby smooth out and/or shift the device contours. These effects can also be incorporated into the preceding model of the fabrication process: Smoothing of contours can be considered as an additional blurring effect and can be incorporated into the imaging kernel $h(x, z)$. Similarly, a shift of the contour can be taken into account by adapting the development threshold g_{th}.

Both the kernel $h(x, z)$ and the development threshold g_{th} thus depend in a complicated way on the imaging system, the substrate properties, the resist properties, the development process and probably on the etch process. It is hard to predict the imaging kernel for a certain fabrication process theoretically. It should however be possible to determine at least certain parameters of $h(x, z)$ by comparing the designed and the real contour trajectories for special test structures. Such model parameters would be extremely helpful to predict the fabrication tolerances that have to be expected for a certain fabrication process. It would then be possible to take the resolution limitations into account during the design.

4.3.2 Contour Trajectory Deformation

The deformation of the contour trajectory is described by the local deviation δ, see Fig. 4.6 (a). Introducing Eqs. (4.17) and (4.18) into Eq. (4.20), the implicit equation for the real contour trajectory can be written as

$$\iint\limits_{D_{\text{exp}}} h(x - \xi, z - \eta)\, \mathrm{d}\xi\, \mathrm{d}\eta - g_{\text{th}} = 0. \tag{4.21}$$

Since the exposed domain D_{exp} can have a very complicated shape, it is in general not possible to solve the integral in Eq. (4.21).

In most cases of practical interest, however, the imaging kernel $h(x, z)$ is well localized in space, and for a certain point (x, z) we can approximate the integral in Eq. (4.21) by only considering the contributions from the direct neighbourhood of this point. For this purpose, we use a second-order approximation of the designed contour in the point $\mathbf{r}(t_0)$, and we switch to local coordinates (u, v). In the (u, v)-system, the second-order approximation of the contour curve is a parabola defined by $v = \frac{\kappa(t_0)}{2}u^2$, where κ is the local curvature of the contour,

$$\kappa(t) = \frac{\frac{\mathrm{d}x}{\mathrm{d}t}\frac{\mathrm{d}^2z}{\mathrm{d}t^2} - \frac{\mathrm{d}^2x}{\mathrm{d}t^2}\frac{\mathrm{d}z}{\mathrm{d}t}}{\left(\frac{\mathrm{d}x}{\mathrm{d}t} + \frac{\mathrm{d}z}{\mathrm{d}t}\right)^{3/2}}.$$

(4.22)

For a negative-tone process, Eq. (4.21) can now be approximated by

$$\int\limits_{\xi=-\infty}^{+\infty} \int\limits_{\eta=\frac{1}{2}\kappa\xi^2}^{+\infty} \overline{h}(u-\xi, v-\eta)\,\mathrm{d}\xi\,\mathrm{d}\eta - g_{\mathrm{th}} = 0,$$

(4.23)

in the vicinity of the origin ($u = 0, v = 0$). Here, $\overline{h}(u, v)$ is the imaging kernel rotated to the coordinate system (u, v),

$$\overline{h}(u, v) = h(u\sin\varphi - v\cos\varphi, u\cos\varphi + v\sin\varphi).$$

(4.24)

For a spatially well-localized imaging kernel, the upper limit of the inner integral in Eq. (4.23) does not matter and we have set it to $+\infty$, assuming that the exposed domain extends far enough into the $v-$direction. For a positive-tone process, analogous arguments can be used, and the inner integral in Eq. (4.23) extends from $-\infty$ to $\frac{1}{2}\kappa\xi^2$.

The deviation δ is now defined by the intersection of the deformed contour with the v-axis. From Eq. (4.23) we find an implicit equation of the from $p(\kappa, \delta) = 0$ connecting the local deviation δ and the local curvature κ,

$$p(\kappa, \delta) := \int\limits_{\xi=-\infty}^{+\infty} \int\limits_{\eta=\frac{1}{2}\kappa\xi^2}^{+\infty} \overline{h}(-\xi, \delta - \eta)\,\mathrm{d}\xi\,\mathrm{d}\eta - g_{\mathrm{th}} = 0.$$

(4.25)

For a well-developed fabrication process without any dimensional bias, straight sections of the contour do not show any deviation, i.e., for $\kappa = 0$ we have $\delta = 0$. This leads to[2]

$$g_{\mathrm{th}} = \int\limits_{\xi=-\infty}^{+\infty} \int\limits_{\eta=0}^{+\infty} \overline{h}(-\xi, -\eta)\,\mathrm{d}\xi\,\mathrm{d}\eta = 0.5,$$

(4.26)

and Eq. (4.25) can be re-written as

$$p(\kappa, \delta) = \int\limits_{\xi=-\infty}^{+\infty} \int\limits_{\eta=\frac{1}{2}\kappa\xi^2}^{0} \overline{h}(-\xi, \delta - \eta)\,\mathrm{d}\xi\,\mathrm{d}\eta = 0.$$

(4.27)

---

[2]Strictly speaking, Eq. (4.26) implies further assumptions as to the shape $h(x, z)$ to guarantee that $\int\limits_{\xi=-\infty}^{\infty} \int\limits_{\eta=0}^{\infty} \overline{h}(-\xi, \delta - \eta)\,\mathrm{d}\xi\,\mathrm{d}\eta$ is independent from the local contour direction φ. Inversion symmetry of $h(x, z)$ with respect to the origin is a sufficient condition.

If the imaging kernel is known, it is possible to calculate the exact relation $\delta = \delta(\kappa)$ from Eq. (4.27). But even without knowing $\overline{h}(u, v)$ exactly, we can approximate the function $\delta(\kappa)$ linearly by means of implicit differentiation, assuming that $\frac{\partial p(\kappa,\delta)}{\partial \delta} \neq 0$,

$$\frac{\mathrm{d}\delta(\kappa)}{\mathrm{d}\kappa} = -\frac{\frac{\partial p(\kappa,\delta)}{\partial \kappa}}{\frac{\partial p(\kappa,\delta)}{\partial \delta}}. \tag{4.28}$$

We find a simple relationship

$$\left.\frac{\mathrm{d}\delta(\kappa)}{\mathrm{d}\kappa}\right|_{\delta=0} = \frac{1}{2}\sigma_u^2, \tag{4.29}$$

where σ_u^2 is the second moment of the imaging kernel along the tangential direction of the contour,

$$\sigma_u^2 = \int\limits_{u=-\infty}^{+\infty} \overline{h}(u, 0)u^2 \, \mathrm{d}u. \tag{4.30}$$

The relationship between the local curvature κ and the local deviation δ can now be approximated to first order,

$$\delta(\kappa) = \frac{1}{2}\sigma_u^2\kappa. \tag{4.31}$$

The result is very intuitive: The contour is always shifted towards the center of the local curvature. The bigger the "width" σ_u of the kernel and the larger the local curvature κ, the stronger the real contour deviates from the designed one. Tight bends and sharp corners are thus very prone to inaccuracies due to limited resolution, and the inaccuracies increase quadratically with σ_u. For a positive-tone process, analogous arguments lead to the same result.

4.3.3 Optimum Contour Trajectories

With the results of the last section it is straightforward to formulate the problem of optimum contour trajectories mathematically. The aggregated deviation error between the designed contour trajectory $C_{\mathrm{cont}}^{(\mathrm{des})}$ and the real contour trajectory $C_{\mathrm{cont}}^{(\mathrm{real})}$ can be rated by an integral of the form $\int_{C^{(\mathrm{des})}_{\mathrm{cont}}} |\delta|^q \, \mathrm{d}s$, $q > 1$, where δ is the local deviation between the curves, $\mathrm{d}s$ is an arc length element of $C_{\mathrm{cont}}^{(\mathrm{des})}$. Using Eq. (4.31), we find that this measure is directly proportional to the integral

$$I - \int\limits_{C_{\mathrm{cont}}^{(\mathrm{des})}} |\kappa|^q \, \mathrm{d}s. \tag{4.32}$$

The goal is now to minimize the value of this integral by finding an optimal contour trajectory $C_{\mathrm{cont}}^{(\mathrm{des})}$. The mathematical problem is identical to the one formulated for the outer boundary of a waveguide bend, Eq. (4.2), and we can directly adopt the solution. Using again the local direction φ as a curve parameter, the ideal class of contour trajectories is given by

$$\mathbf{r}(\varphi) = \mathbf{r}(\varphi_0) + C_1^{-1}\int\limits_{\varphi_0}^{\varphi} \frac{1}{[\cos(\widetilde{\varphi} - C_2) + C_3]^{1/q}} \begin{pmatrix} \sin(\widetilde{\varphi}) \\ \cos(\widetilde{\varphi}) \end{pmatrix} \mathrm{d}\widetilde{\varphi}. \tag{4.33}$$

The angle φ_0 defines the local direction of the curve in the starting point $\mathbf{r}(\varphi_0)$. Fixing the ending point we obtain two equations for the three constants C_1, C_2, C_3,

$$\mathbf{r}(\varphi_1) - \mathbf{r}(\varphi_0) = C_1^{-1} \int\limits_{\varphi_0}^{\varphi_1} \frac{1}{[\cos{(\varphi - C_2)} + C_3]^{1/q}} \left(\begin{array}{c} \sin(\widetilde{\varphi}) \\ \cos(\widetilde{\varphi}) \end{array} \right) \mathrm{d}\,\widetilde{\varphi}, \tag{4.34}$$

where φ_1 defines the direction in the end point. The remaining degree of freedom can be used to adapt the curvature of the contour in the starting point or the ending point.

4.3.4   Numerical Verification

In the last section, we have derived a class of ideal contour trajectories which show maximal stability with respect to fabrication processes of limited resolution. We will now investigate the effects of contour deformation on the waveguide bends introduced in Section 4.2.2. For this purpose, the designed normalized exposure profile $g^{(\mathrm{des})}(x, z)$ is calculated for each bend. As an imaging kernel, we assumed a two-dimensional Gaussian of equal variance σ in the x and the z-direction,

$$h(x, z) = \frac{1}{2\pi\sigma^2} \mathrm{e}^{-\frac{x^2 + z^2}{2\sigma^2}} . \tag{4.35}$$

Figure 4.7 (a) shows the designed normalized exposure profile $g^{(\mathrm{des})}(x, z)$ for a waveguide bend consisting of a semicircular disc and a taper section. The layout of the bend is sketched in Fig. 4.2 (b), and the geometrical parameters are $R_1 = 1.5\,\mu\mathrm{m}$, $R_2 = 1.177\,\mu\mathrm{m}$, $l_{\mathrm{tpr}} = 1.5\,\mu\mathrm{m}$, $w_{\mathrm{ring}} = 0.478\,\mu\mathrm{m}$. Figure 4.7 (b) shows the corresponding real exposure profile $g^{(\mathrm{real})}(x, z)$ for a Gaussian imaging kernel with $\sigma = 0.3\,\mu\mathrm{m}$. The designed and the real contour curve $C_{\mathrm{cont}}^{(\mathrm{des})}$ and $C_{\mathrm{cont}}^{(\mathrm{real})}$ are depicted in Fig. 4.7 (c). It can be seen that curved sections of the contour are shifted towards the center of the curve, and the shift increases with the curvature.

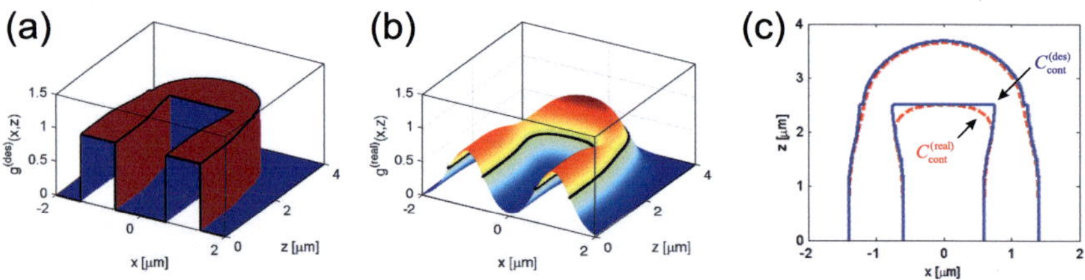

Fig. 4.7. (a) Designed normalized exposure profile $g^{(\mathrm{des})}(x, z)$ for a waveguide bend consisting of a semicircular disc and a taper section similar to Fig. 4.2 (b).
(b) Real exposure profile $g^{(\mathrm{real})}(x, z)$ for a Gaussian imaging kernel with $\sigma = 0.3\,\mu\mathrm{m}$. (c) Designed contour curve $C_{\mathrm{cont}}^{(\mathrm{des})}$ and real contour curve $C_{\mathrm{cont}}^{(\mathrm{real})}$.

We have applied this procedure to different bend types similar to those shown in Fig. 4.7, but optimized for $R_1 = 1.0\,\mu\mathrm{m}$. For each bend, we have determined the insertion

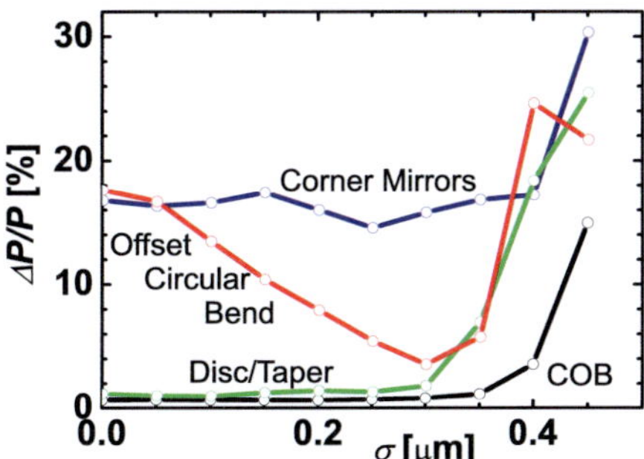

Fig. 4.8. Normalized power loss vs. σ for the contour optimized bend (COB), the disc/taper configuration, the laterally offset circular bend, and the pair of corner mirrors.

loss for different values of σ by 2D FDTD simulation. The result is plotted in Fig. 4.8. As expected, the COB is very robust with respect to the resolution of the fabrication process, and performance only drops for $\sigma \geq 0.4\,\mu$m, which corresponds to a full width at half the maximum (FWHM) of the imaging kernel of $0.94\,\mu$m. The disc/taper configuration keeps its high performance for low values of σ, but starts degrading for $\sigma \geq 0.3\,\mu$m. For the offset circular bend and the corner mirrors, the insertion loss can even decrease if σ increases, i. e., if the corners in the outer bend contours are smoothed out. This indicates that sharp corners in the outer contour of a bent waveguide yield designs that are ab initio suboptimal.

The ideal contour trajectories derived in this section can not only be applied to waveguide bends, but also for the design of much more complicated structures. For the design of multi-mode interference (MMI) couplers, the rectangular multi-mode section could, e.g., be replaced by a smooth shape defined by a set of model points that are smoothly linked by contours according to Eq. (4.33). Numerical optimization can then be used to find the ideal position of the model points, while maintaining ideal contours that are robust with respect to the resolution of the fabrication process.

4.4   Summary

For ultracompact bends in singlemode-operated multimode waveguides we derived analytically a class of outer contour trajectories with minimum loss and maximum stability with respect to fabrication processes of limited resolution. 2D FDTD is used to numerically optimize the contour parameters for single-mode operation, resulting in broadband COB with low insertion loss and mode conversion. The performance is maintained even for low-resolution fabrication processes with an imaging kernel of more than 500 nm FWHM. 3D FDTD modelling predicts losses of $< 0.1\,$dB $(< 0.2\,$dB$)$ over a 100 nm (300 nm) bandwidth for a 180° bend with $1.5\,\mu$m radius. The simulations are supported by measuring a roughness-limited loss of 0.35 dB.

Chapter 5

Minimizing Waveguide Loss by Technology: Preferential Wet Etching

5.1 Introduction

In Chapter 3 we have derived design guidelines for low-loss straight waveguides affected by sidewall roughness: They should be wide to reduce the strength of the electric field at the rough boundary and to maintain a large propagation constant, and they should be flat to minimize the scattering surface. For bent waveguide sections, or, more general, for bent contours of planar lightwave circuits (PLC), we have derived ideal contour trajectories in Chapter 4. Based on this information, low-loss passive PLC can be designed.

There are, however, cases of high index-contrast (HIC) waveguides to which the aforementioned design guidelines cannot be applied. As discussed in more detail in Chapters 6 and 7, the interaction of the guided light with a low-index cover material can be of considerable technical interest. This interaction can be enhanced by exploiting the discontinuity of the normal electric field component at the surface of the high index waveguide core. The enhancement is particularly strong if the cover material is deposited into a groove, along which the light is guided, see Figs. 6.1, 6.2, and 7.1 (b). For these configurations, as a matter of fact, high electric fields occur in the vicinity of the rough waveguide sidewalls. Hence, reducing the field strength at the waveguide sidewalls is only possible at the expense of decreasing the interaction with the cover material. It is therefore of prime importance to fabricate these waveguides as smooth as possible. Fortunately, the geometries of such strip and slot waveguide configurations are rather simple: Straight waveguide sections and grooves with polygonal cross sections or cavities with polyhedral shapes are sufficient.

In this chapter, we show that such waveguide structures can be fabricated with atomically smooth surfaces by preferential wet etching of crystalline materials. We present a self-aligning fabrication process which allows to embed such waveguide structures into conventional PLC, thereby combining the geometrical flexibility of structures fabricated by anisotropic (dry) etching processes with the outstanding precision of waveguides obtained by preferential (wet) etching. We discuss in detail the waveguide geometries that can be obtained in the crystalline device layer of an silicon-on-insulator (SOI) wafer. For $\langle 100 \rangle$-oriented device layers, we have fabricated promising prototype structures.

This chapter is structured as follows: In Section 2 we give an overview of preferential etching of semiconductor materials, with an emphasis on silicon. In Section 3 we then discuss different waveguide geometries that can be obtained by preferential etching of crystalline SOI device layers. We then present two examples of self-aligning fabrication processes that allow to combine preferential etching techniques with conventional anisotropic dry etching, see Section 4. In Section 5, we report on the fabrication of prototype structures. We discuss technological details and we investigate the effects of misalignment between the structure and the crystallographic directions.

5.2   Preferential Etching of Semiconductors

To avoid confusion, let us first define some fundamental terms that will be used in this chapter:

Preferential etching In the following the term "preferential etching" refers to an etching process occurring preferentially along certain crystallographic directions, i.e., different crystal planes are etched with significantly different rates. Planes with low etch rates shall be referred to as stable crystal planes. Etch rates for stable planes are typically more than ten times smaller than for other crystal planes.

Anisotropic etching The term "anisotropic etching" denotes an etching process for which the etch rate in the vertical direction (normal to the wafer surface) is much higher than in the horizontal directions. Anisotropic etching processes are normally based on dry etching techniques that are optimized for vertical sidewalls and minimal undercutting, e.g., RIE, inductively-coupled plasma reactive ion etching (ICP-RIE), ion beam etching (IBE), chemically-assisted ion beam etching (CAIBE), or reactive ion beam etching (RIBE).

Isotropic etching "Isotropic etching" refers to an etching process for which the etch rate does not depend on the direction, e.g., the dissolution of silicon dioxide (SiO_2) in concentrated hydrofluoric acid (HF).

Material-selective etching All etching processes can further be material-selective, i.e., they dissolve certain materials while leaving others basically unaffected. Material selective processes can, e.g., be used to remove certain parts of etch masks.

Preferential etching of semiconductor materials is usually based on wet etching, and we will review different etching techniques in the following sections. For silicon, such techniques are routinely employed for the fabrication of micro-electro-mechanical systems (MEMS) [85]. For III-V compound semiconductors, preferential wet etching techniques exist, but play only a minor role.

In the following, crystallographic directions are specified by Miller indices. A triple of integers $[lmn]$ denotes a direction in the basis of the lattice vectors, and $\langle lmn \rangle$ comprises all directions that are equivalent to $[lmn]$ by symmetry of the crystal. Accordingly, (lmn)

denotes the plane with normal direction $[lmn]$, and $\{lmn\}$ is the set of planes that are equivalent to (lmn) by symmetry of the crystal. Negative values of indices are denoted by a bar, e.g., $\bar{1}$ for -1.

5.2.1 Preferential Wet Etching of Crystalline Silicon

While preferential wet etching of silicon has been widely used for the fabrication of micro-electro-mechanical system (MEMS), reports on fabrication of optical waveguides are limited to the reduction of surface roughness of pre-structured waveguides [63], or to the etching of V-grooves as templates for polymer waveguides [10]. Special nanophotonic waveguide geometries have neither been proposed nor have prototypes been fabricated, and it is unclear whether the precision of preferential wet etching methods developed for micrometer-scale MEMS can be transferred to nanometer-scale optical waveguides. For structures with high aspect-ratio trenches, the etch rate at the bottom of the trench might be limited by diffusion, leading to incomplete etching of the profile.

Diamond Lattice and Crystal Planes

In crystalline form, silicon atoms are connected to each other by strong covalent bonds of $3sp^3$-hybridized orbitals, thus forming a diamond lattice. Each atom features four bonds that are directed to the corners of a tetrahedron, hence defining angles of $109.47\,°$. Fig. 5.1 (a) shows the cubic unit cell. The basic lattice vectors correspond to the edges of the sketched cube. Due to full inversion and permutation symmetry with respect to the basic lattice vectors, the set of crystallographic directions $\langle lmn \rangle$ comprises all directions $[lmn]$ that can be obtained by inverting and/or permuting any of the indices l, m, or n. The same applies to the set of crystallographic planes $\{lmn\}$.

Fig. 5.1 (b) shows the most important crystallographic planes: The $\{100\}$ planes are perpendicular to the basic lattice vectors, the $\{110\}$ planes are perpendicular to the diagonals of the faces, and $\{111\}$ denotes the set of planes that are perpendicular to the space diagonals. Fig. 5.1 (c) shows the orientation of the covalent atomic bonds with respect to the $\{100\}$ and $\{111\}$ planes.

Chemical Reaction

Aqueous alkaline solutions can be used as preferential wet etchants for crystalline silicon. In this reaction, a $[SiO_2(OH)_2]^{2-}$ complex is formed, which has been verified by Raman spectroscopy [82]. The overall reaction is

$$Si + 2OH^- + 2H_2O \rightarrow [SiO_2(OH)_2]^{2-} + 2H_2. \qquad (5.1)$$

When depicted as a function of the etchant's molarity, the etch rates for $\{100\}$ and $\{110\}$ show a maximum, which has been explained by hydration effects [38].

The electrochemical reaction mechanisms are still under debate in the literature. A two-step electrochemical reaction mechanism is proposed in [94]: By a reaction with hydroxide ions, a silicon atom is removed from the solid surface, and four electrons are injected into the conduction band [94]. This corresponds to an oxidation of silicon,

$$Si + 2OH^- \rightarrow [Si(OH)_2]^{2+} + 4e^-. \qquad (5.2)$$

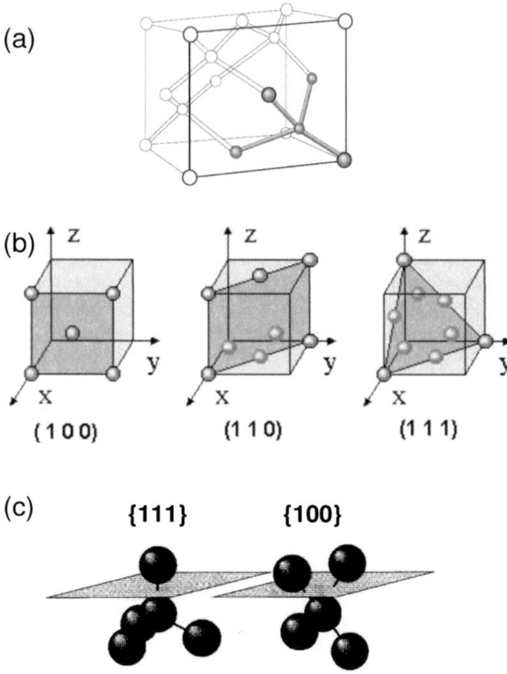

Fig. 5.1. Crystalline structure of silicon; (a) Unit cell of the silicon diamond lattice; (b) Important crystallographic planes; (c) Orientation of atomic bonds with respect to the {111}- and the {100}-planes (modified after [75])

The free electrons are localized near the solid surface due to downward bending of the energy bands [94]. They can then be transferred to water molecules, reducing them to hydroxide ions and gaseous hydrogen,

$$4H_2O + 4e^- \rightarrow 4OH^- + 2H_2 \uparrow . \tag{5.3}$$

The complex $[Si(OH)_2]^{2+}$ instantaneously reacts with the hydroxide ions, to form the water soluble $[SiO_2(OH)_2]^{2-}$ complex,

$$[Si(OH)_2]^{2+} + 4OH^- \rightarrow [SiO_2(OH)_2]^{2-} + 2H_2O. \tag{5.4}$$

Preferential Etching

The etch rates for the $\langle 111 \rangle$ directions can be more than 100 times smaller than for the $\langle 100 \rangle$ and $\langle 110 \rangle$ directions [85, 94]. Different models of the reduced etch rates for {111} surfaces have been published, and possible explanations involve specific bond configurations on different crystal planes [125] and blocking of the reaction on {111} surfaces by OH:3H$_2$O or other threefold complexes [55].

A simple qualitative explanation of the chemical stability of {111} surfaces can be given on the basis of Fig. 5.1 (c): Each atom on a {111} surface is linked to the bulk crystal by three bonds and only possesses one "dangling bond" to which the solvent can attach. It is therefore much more "difficult" for the hydroxide ions to remove a {111} surface atom than a {100} surface atom, which is held by only two crystal bonds and has two dangling bonds. The etch rate for the {100} surfaces is hence much higher than for the {111} surfaces. For the {110} surfaces (not depicted in Fig. 5.1 (c)), each atom has one dangling bond, one bond into the bulk and two bonds in the surface plane. The bonds in the surface plane can be easily cracked by the hydroxide ions, resulting in a high etch rate also for the {110} surfaces [55].

Etchants and Etch Rates

Preferential wet etchants are aqueous alkaline solutions, where the main component can either be organic or inorganic. The etch rates and the degree of anisotropy strongly depends on the etchant. Unlike applications in MEMS fabrication, where deeply etched structures in silicon substrates are created, etch depths for optical applications are rather small. It is hence advantageous to work at low or moderate etch rates between 5 nm / min and 1 000 nm / min, and selectivity requirements with respect to mask materials are relaxed. On the other hand, the surface quality of the stable crystalline plane is much more important for optical devices than for MEMS.

As for most chemical reactions, the temperature dependence of the reaction rate R (etch rate) for a certain etchant and a certain crystallographic direction follows from the Arrhenius equation,

$$R(T) = R_0 \exp\left(-\frac{E_a}{kT}\right). \tag{5.5}$$

T denotes the absolute temperature, and $k = 1.280658 \times 10^{-23}$ W s / K is Boltzmann's constant. The prefactor R_0 and the activation energy E_a are determined from measuring the etch rates at different temperatures. In the following, crystal-dependent etch rates for {lmn} surfaces are denoted as R_{lmn}.

To avoid geometrical inaccuracies due to mask erosion, the selectivity of the etchant with respect to the mask material should be high. A summary of etch rates for different etchants and mask materials can be found in [58] and [119]. In the next sections, we will review the most widely used etchants, estimate the etch rates, the degree of crystallographic anisotropy, and the selectivity with respect to different mask materials. We will also discuss practical issues like compatibility with complementary metal-oxide-semiconductor (CMOS) technology.

Aqueous Alkali Hydroxide Solutions Among the inorganic preferential etchants, the most frequently used is an aqueous solution of potassium hydroxide (KOH). At room temperature, the etch rates peak at a concentration of approximately 20 % of KOH by mass [38]. Numerical values of R_0 and E_a for the $\langle 100 \rangle$ and the $\langle 110 \rangle$ directions and for

Etchant	T	Si Etch Rates [nm / min]		Anisotropy Ratios		Mask Etch Rates [nm / min]		Ref.
		R_{110}	R_{100}	$\frac{R_{110}}{R_{100}}$	$\frac{R_{100}}{R_{111}}$	R_{SiO_2}	$R_{Si_3N_4}$	
KOH, 20 %	25 °C	57	39	1.5	~ 100	0.026	–	[94]
KOH, 20 %	40 °C	180	120	1.5	~ 100	0.125	–	[94]
KOH, 45 %	25 °C	37	25	1.5	~ 300	0.035	–	[94, 54]
KOH, 45 %	40 °C	115	77	1.5	~ 300	0.173	–	[94, 54]
EDP, type "S"	20 °C	40	20	2	~ 150	–	–	[94]
EDP, type "S"	50 °C	137	89	1.5	~ 100	–	–	[94]
TMAH, 40 %	20 °C	55^e	1.3^e	42^e	$\sim 8^e$	$< 0.01^e$	–	[105, 100]
TMAH, 40 %	50 °C	206	21	9.8	~ 40	< 0.01	–	[105]
TMAH, 40 %	70 °C	440	104	4.2	~ 60	~ 0.02	–	[105]

Table 5.1. Preferential wet etching processes for the fabrication of silicon waveguide nanostructures. The considered etchants comprise aqueous solutions of KOH, Ethylenediamine and Pyrocatechol (EDP), and Tetramethylammonium hydroxide (TMAH). The concentrations are specified in percent by mass. T denotes the temperature of the solution, and R_{lmn} stands for the rate with which $\{lmn\}$ planes are etched in silicon. The ratios R_{100}/R_{111} specify the crystallographic anisotropy of the etch rates. R_{SiO_2} and $R_{Si_3N_4}$ denote the etch rates for silicon dioxide and silicon nitride. Dashes (—) are used where the etch rates were too low to be measured. The corresponding literature references are given in the last column. For the 40% TMAH solution at 20°C, data had to be extrapolated from higher temperatures by means of Eq. (5.5), and the accuracy may therefore be limited. This is indicated by a superscript "e".

different concentrations and can be found in [94, Tab. II]. Favoured concentrations of KOH range from 10% to 60% by mass, and the preferred temperature of the etchant is between 20 °C and 70 °C. Etch rates for the $\langle 100 \rangle$ directions then lie between 8 nm / min to 27 nm / min at 20 °C and 250 nm / min to 820 nm / min at 70 °C. For the $\langle 110 \rangle$ direction, the corresponding ranges are 12 nm / min … 38 nm / min at 20 °C and 400 nm / min … 1200 nm / min at 70 °C [94, Tables A-I and A-II]. Higher etch rates are not of interest for optical waveguide structures. For an unspecified KOH concentration, the ratio R_{110} : R_{100} : R_{111} of the etch rates for the main crystal planes $\{110\}$, $\{100\}$, and $\{111\}$ was reported to range from 50 : 30 : 1 at 100 °C to about 160 : 100 : 1 at room temperature [94]. An extraordinarily high ratio of 600 : 300 : 1 was reported for an aqueous solution containing 44 % KOH by mass [54]. Concerning mask selectivity, silicon nitride (Si_3N_4) does not show measurable etch in KOH and seems to be more resistant than silicon dioxide ($R_{SiO_2} = 0.007$ nm / min to 3.25 nm / min) [119], [94, Tab. A-III].

For the fabrication of optical waveguides, working at low temperatures between 20 °C and 40 °C has the advantage of high anisotropy ratio R_{110} : R_{100} : R_{111}, while still preserving sufficiently large etch rates. At the same time, underetching of the mask is less crucial. Table 5.1 lists data for four examples of KOH-based wet etch processes that can be used for the fabrication of waveguide structures.

Unfortunately, potassium ions can be detrimental to CMOS integrated circuits, and KOH based etchants are therefore not compatible with CMOS technology [58]. However, there are hydroxide-based etchants that incorporate organic compounds rather than alkali ions, and are thus fully CMOS compatible. Two important examples shall be reviewed in the following sections.

Aqueous Solutions of Ethylenediamine and Pyrocatechol (EDP): Unlike KOH-based etchants, EDP solutions consisting of water, ethylenediamine (1,2-diaminoethane, $C_2H_8N_2$) and pyrocatechol (benzene-1,2-diol, $C_6H_6O_2$) do not contain alkali ions. They are therefore CMOS compatible, yet at the expense of being corrosive and potentially carcinogenic [58].

For slow etching of silicon at lower temperatures, an optimized composition of oxygen-resistant EDP solution has been developed [88]. This type "S" etchant consists of a mixture of 133 ml of water, 1000 ml of ethylenediamine, 160 g of pyrocatechol, and 6 g of pyrazine ($C_4H_4N_2$). For temperatures between 20 °C and 70 °C, etch rates along the $\langle 110 \rangle$ and the $\langle 100 \rangle$ directions range from 40 nm / min to 270 nm / min and from 20 nm / min to 200 nm / min, respectively [94, Fig. 13]. Anisotropy ratios $R_{\langle 110 \rangle}$: $R_{\langle 100 \rangle}$: $R_{\langle 111 \rangle}$ range from 300 : 150 : 1 at 20 °C to 100 : 75 : 1 at 70 °C, and are thus comparable to the ones for KOH [94, Fig. 13]. At 70 °C the etch rate R_{SiO_2} for SiO_2 is approximately 0.08 nm / min, and at 20 °C R_{SiO_2} is negligible (< 0.001 nm / min) [94, Fig. 18]. Etch rates for Si_3N_4 seem to be too small to be measured [94, 58].

Residues on the etched surface may form if large amounts of silicon are dissolved in type "S" EDP solutions at temperatures below 50 °C [88]. However, for etching structures with sub-micrometer depths, only diminutive amounts of silicon are dissolved, and saturation of the etchant will not be a severe problem. In Table 5.1 we give two typical examples of EDP-based preferential etching processes that lend themselves for the fabrication of silicon waveguide structures.

Aqueous Solutions of Quaternary Ammonium Hydroxides Tetramethylammonium hydroxide (TMAH, $(CH_3)_4OH$) dissolved in water is a useful preferential etchant for silicon. TMAH is CMOS compatible and considered less harmful than EDP [58]. Being constituent of most metal-ion-free (MIF) developers, highly pure, VLSI-grade TMAH is available in most cleanrooms.

For increasing TMAH concentrations above 5 % by mass, the etch rates for the $\{100\}$ and the $\{110\}$ planes and the anisotropy ratio R_{100} : R_{111} decrease steadily [105, Figs. 1, 3, 5]. It is therefore advantageous to work at high concentrations. The etch rates can then be controlled by adjusting the temperature. For a TMAH concentration of 40 % by mass and temperatures between 60 °C and 90 °C, R_{100} : R_{111} ranges from 25 : 1 to 50 : 1 [105, Fig. 5]. At the same time according to [105], the etch rates along the $\langle 110 \rangle$ and the $\langle 100 \rangle$ directions also differ vastly: For temperatures between 20 °C and 80 °C, R_{110} ranges from 55 nm / min to 620 nm / min, whereas R_{100} only changes between 1.3 nm / min to 216 nm / min. For the 40 % TMAH solution at 20 °C, data had to be extrapolated from higher temperatures by means of Eq. 5.5. The data marked with a superscript "e" may therefore be of limited accuracy. It can nevertheless be observed, that for low temperatures $T \leq 50$ °C preferential etching occurs also with respect to the $\{110\}$ and the $\{100\}$ planes, whereby the $\{100\}$ planes can be considered stable (R_{110} : $R_{100} \geq 10$). This tendency is qualitatively confirmed by other authors, but with lower R_{110} : R_{100} ratio of approximately 5 [100]. Still, investigating the $\{110\}/\{100\}$-anisotropy could be worthwhile since it may considerably increase the diversity of waveguide geometries that can be fabricated by preferential wet etching.

Below $80\,^\circ$C, the etch rates for SiO_2 are less than $0.05\,\text{nm}\,/\,\text{min}$ [105, Fig. 6], and for Si_3N_4, the etch rate seems to be negligible [105]. Selectivity with respect to common mask materials is hence fully sufficient.

5.2.2 Preferential Wet Etching of III-V Compound Semiconductors

Most III-V compound semiconductors have zincblende crystal lattices[1]. As a consequence, $\{111\}$ surfaces can either be covered by group-III or group-V atoms. The former surfaces are commonly referred to as (111) (or $\{111\}$A), whereas the latter are labelled with $(\bar{1}\,\bar{1}\,\bar{1})$ (or $\{111\}$B) [113, 53]. Each $\{111\}$ surface atom has three remaining bonds to the solid. For the (111) surfaces made up of group-III atoms, all outer electrons are consumed in these bonds, whereas group-V atoms on a $(\bar{1}\,\bar{1}\,\bar{1})$ surface have two spare electrons that can take part in a reaction. It is therefore to be expected that etching of $(\bar{1}\,\bar{1}\,\bar{1})$ planes is much faster than of (111) planes.

Preferential wet etching techniques for III-V compound semiconductors are far less sophisticated than for silicon, but may still be useful for simple waveguide structures. For GaAs, (111) planes exhibit stable behavior when etched with a solution of bromine (Br_2) and methanol (CH_3OH). In contrast to that, the $(\bar{1}\,\bar{1}\,\bar{1})$ surfaces are etched as quickly as other crystal planes, $R_{111} \ll R_{100} < R_{\overline{111}} < R_{110}$, where $R_{111} : R_{100}$ ratios are of the order of 5 [107]. Similar etchants have been used to fabricated high-quality V-grooves with (111) sidewalls in InP substrates [53]. Mixtures of hydrochloric acid (HCl) and nitric acid (HNO_3) can also be used for preferential etching of InP, $R_{111} \ll R_{100} < R_{\overline{1}\,\overline{1}\,\overline{1}}$ and $R_{111} : R_{100} \approx 15$ [113]. An overview over different etchants for compound semiconductors can be found in [112].

5.3 Crystal-Aligned Waveguides and Cavities in SOI

In this section we present special waveguide geometries that can fabricated with high precision in the device layer of a silicon-on-insulator (SOI) wafer using preferential wet etching techniques. SOI wafers consist of a silicon handle wafer, covered with a layer of silicon dioxide (buried oxide, BOX) and with a silicon device layer (DEV). For optical applications, the BOX must be thick enough (typically $> 1\,\mu$m) to prevent coupling of light into the substrate. Typical device layer thicknesses range from $150\,$nm to $500\,$nm. SOI wafers with thin high-quality crystalline device layers can, e.g., be fabricated by deploying a so-called SMART CUT^{TM} process [15]. SMART CUT^{TM} is Soitec's proprietary technology, used to manufacture so-called UNIBOND SOI wafers (http://www.soitec.com/en/techno/t_2.htm).

When fabricated with preferential wet etching, the shape of the waveguides must be adapted to the crystalline structure of the silicon device layer: The sidewalls have to be plane and must coincide with stable $\{111\}$ planes, and any convex corners must be avoided. Depending on the crystal orientation of the SOI device layer, this leads to different geometries. In the following, the crystal orientation of the device layer is specified

---

[1]An exception is, e. g., GaN, which has wurtzite crystal structure.

by the crystallographic direction $\langle lmn \rangle$, which is associated with the surface normal. The surface of the wafer is thus a $\{lmn\}$ plane.

For certain geometries, an intermediate dry etching step is necessary. The surface of a $\langle 111 \rangle$-oriented device layer is, e.g., a stable plane that is not attacked by the discussed preferential wet etchants. It can only be structured if a dry etching step is performed first. For $\langle 100 \rangle$-oriented device layers, special geometries can be obtained by using a combination of anisotropic dry etching and preferential wet etching.

$\langle 100 \rangle$-Oriented Device Layer

Figure 5.2 depicts waveguide cross sections that can be fabricated from a $\langle 100 \rangle$-oriented SOI device layer. The z-axis (parallel to the waveguide axis) is oriented along a $\langle 110 \rangle$ direction of the crystal. The shapes shown in Figs. 5.2 (a) and (b) can be obtained by directly exposing the $\{100\}$ surface to a preferential etchant which stops at $\{111\}$ planes. To fabricate the cross sections depicted in Figs. 5.2 (c), an anisotropic dry etching step is performed first to generate the trapezoidal cross section $AEDB$. The sidewalls are then etched back to the points F and C by subsequent wet etching. The slot waveguide in Fig. 5.2 (d) is obtained in the same way. It is impossible to fabricate such cross sections

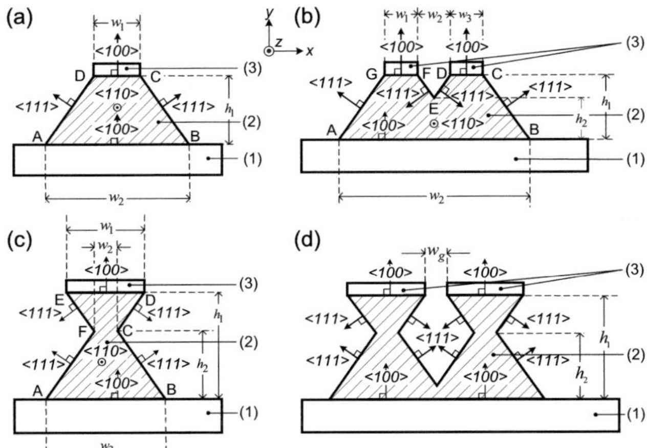

Fig. 5.2. Cross sections of waveguides that can be fabricated in a $\langle 100 \rangle$-oriented SOI device layer by preferential wet etching. For (c) and (d), an anisotropic etching step is followed by preferential wet etching. (1) Silicon dioxide buffer layer, (2) Silicon waveguide core, (3) Etch mask; (a) Trapezoidal cross section, $\angle CBA = \angle BAD \approx 54.74°$ and $\angle ADC = \angle DCB \approx 125.26°$; (b) Heptagonal cross section, $\angle BAG = \angle CBA \approx 54.74°$, $\angle AGF = \angle GFE = \angle EDC = \angle DCB \approx 125.26°$ and $\angle DEF \approx 70.53°$; (c) Hexagonal cross section, $\angle CBA = \angle EDC = \angle FED = \angle BAF \approx 54.74°$ and $\angle BCD = \angle EFA \approx 109.47°$; (d) Polygonal cross section with groove; For (d), the angles and crystallographic directions follow from (a) (b) and (c) by analogy.

with conventional dry etching techniques, not to mention the outstanding precision that can be obtained by preferential wet etching.

Figures 5.3 and 5.4 show cavities that can be fabricated in $\langle 100 \rangle$-oriented SOI device layers. Direct preferential wet etching leads to the cavity depicted in Fig. 5.3, whereas combination with a dry etching step leads to Fig. 5.4. A regular arrangement of such high-precision cavities can exhibit a photonic band gap photonic band gap (PBG) that inhibits propagation of light in the device layer and can hence be used to define photonic crystal (PhC) structures.

$\langle 110 \rangle$-Oriented Device Layer

Figure 5.5 depicts waveguide cross sections that can be fabricated from a $\langle 110 \rangle$-oriented SOI device layer. The z-axis is oriented along a $\langle 112 \rangle$ direction of the crystal. Preferential wet etching techniques can either be used to created the waveguides from the bulk SOI layer, or they can be used after a dry etch process to polish the sidewalls and to make them vertical. It is difficult to fabricate deep and narrow trenches as depicted in Figs. 5.5 (c) and (d) with smooth sidewalls by conventional dry etching techniques. In the cases of Figs 5.5 (b) a thin silicon slab (5) of the device layer remains on the silicon dioxide buffer. This slab region can be used to electrically contact the waveguide core. The same holds true for the slot waveguide structure in Fig 5.5 (d), which, however, requires a two-step process to etch the slot down to the oxide while leaving silicon slabs to the outer sides of the ribs.

Figure 5.6 shows a cavity that can be fabricated in a $\langle 110 \rangle$-oriented device layer and which can be used to define PhC structures.

$\langle 111 \rangle$-Oriented Device Layer

Figures 5.7 and 5.8 depict waveguide cross sections and cavities that can be fabricated from a $\langle 111 \rangle$-oriented SOI device layer. The z-axis of the waveguides is oriented along a $\langle 110 \rangle$ direction of the crystal. Since the surface of the device layer is a stable $\{111\}$ plane, prestructuring by dry etching is necessary. For the waveguide depicted in Fig. 5.7 (a), this prestructuring leads to a the trapezoidal cross section $AD'CB'$. Preferential wet etching techniques can then be used to generate the parallelogram cross sections $ABCD$ of the waveguides. It is impossible to fabricate inclined deep and narrow trenches as depicted in Figs. 5.7 (c) and (d) with smooth sidewalls by conventional dry etching techniques. In the cases of Figs 5.7 (b) and (d), thin silicon slabs (5) of the device layer remain on the silicon dioxide buffer. The slot waveguide depicted Fig 5.7 (d) hence requires again a two-step dry etching process. Similarly, the cavity depicted in Fig. 5.8 is obtained by dry etching of a hexagonal hole followed by preferential wet etching.

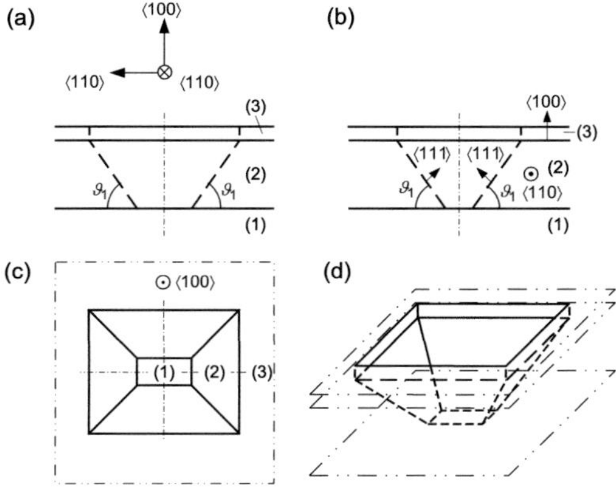

Fig. 5.3. Open cavity that can be fabricated in a $\langle 100 \rangle$-oriented SOI device layer by preferential wet etching; $\vartheta_1 = 54.74\,°$; (1) Silicon dioxide buffer layer, (2) Silicon waveguide core, (3) Etch mask; (a), (b) Upright projections; (c) Horizontal projection; (d) Perspective projection

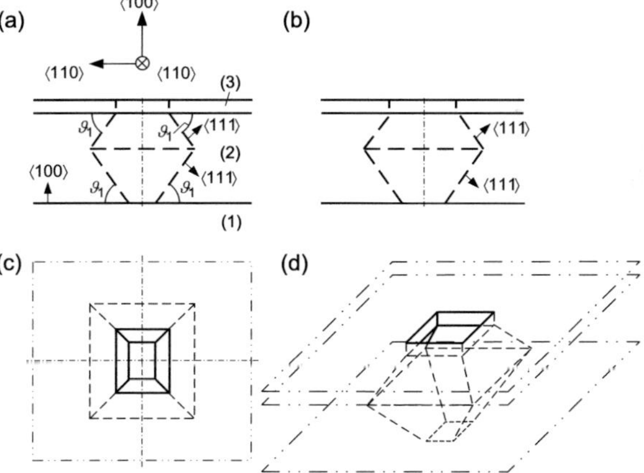

Fig. 5.4. Open cavity that can be fabricated in a $\langle 100 \rangle$-oriented SOI device layer. An anisotropic etching step is followed by preferential wet etching. $\vartheta_1 = 54.74\,°$; (1) Silicon dioxide buffer layer, (2) Silicon waveguide core, (3) Etch mask; (a), (b) Upright projections; (c) Horizontal projection; (d) Perspective projection

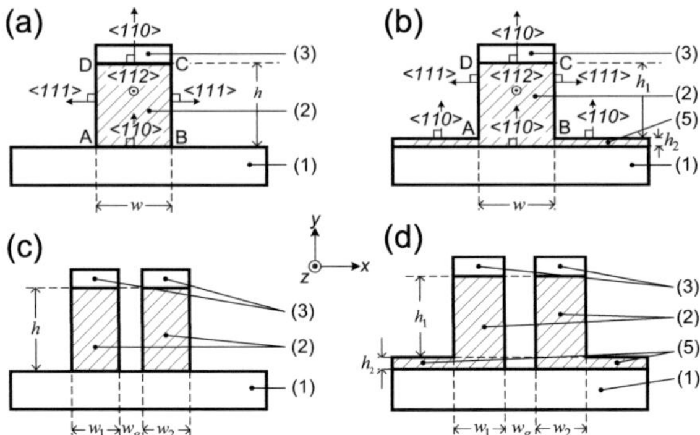

Fig. 5.5. Cross sections of waveguides that can be fabricated in a $\langle 110 \rangle$-oriented SOI device layer; (1) Silicon dioxide buffer layer, (2) Silicon waveguide core, (3) Etch mask, (5) Slab region; (a) Strip waveguide with rectangular cross section; (b) Strip waveguide with rectangular cross section and slab region; (c) Slot waveguide; (d) Slot waveguide with slab regions. For (c) and (d), the crystallographic orientations follow from (a) and (b) by analogy.

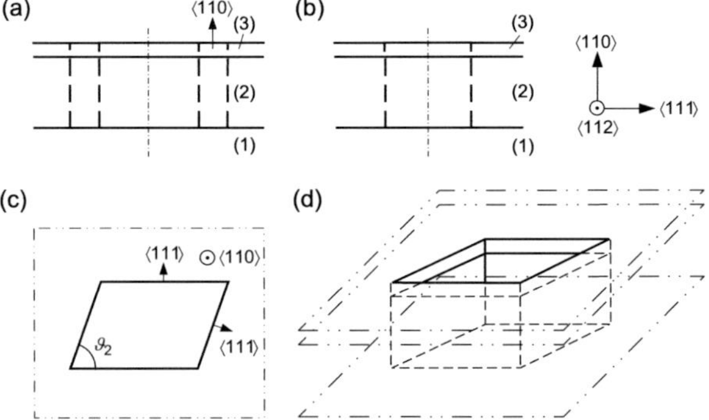

Fig. 5.6. Open cavity that can be fabricated in a $\langle 110 \rangle$-oriented SOI device layer by preferential wet etching; $\vartheta_2 = 70.53\,°$; (1) Silicon dioxide buffer layer, (2) Silicon waveguide core, (3) Etch mask; (a), (b) Upright projections; (c) Horizontal projection; (d) Perspective projection

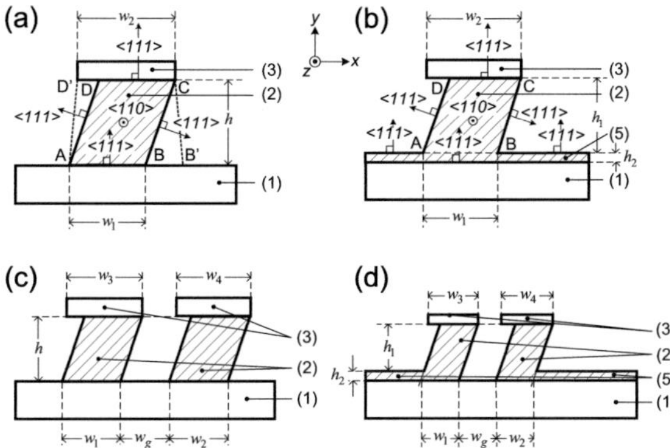

Fig. 5.7. Cross sections of waveguides that can be fabricated in a $\langle 111 \rangle$-oriented SOI device layer. An anisotropic etching step is followed by preferential wet etching. (1) Silicon dioxide buffer layer, (2) Silicon waveguide core, (3) Etch mask, (5) Slab region; (a) Strip waveguide with parallelogram-shaped cross section, $\angle CBA = \angle ADC \approx 109.47\,^\circ$, $\angle BAD = \angle DCB \approx 70.53\,^\circ$; (b) Strip waveguide with slab region; (c) Slot waveguide; (d) Slot waveguide with slab regions. For (c) and (d), the angles and crystallographic orientations follow from (a) and (b) by analogy.

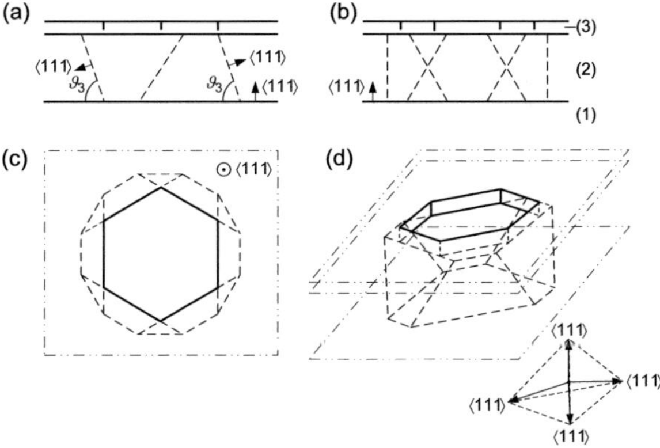

Fig. 5.8. Open cavity that can be fabricated in a $\langle 111 \rangle$-oriented SOI device layer. An anisotropic etching step is followed by preferential wet etching. $\vartheta_3 = 70.53\,^\circ$; (1) Silicon dioxide buffer layer, (2) Silicon waveguide core, (3) Etch mask; (a), (b) Upright projections; (c) Horizontal projection; (d) Perspective projection

5.4   Embedding of Crystal-Aligned Structures into Conventional Planar Lightwave Circuits

As discussed in the previous section, waveguide structures fabricated by preferential etching must be aligned with the crystal lattice and are hence subject to geometrical restrictions which make the realization of basic waveguide elements (e.g., bends) impossible. To overcome these restrictions, it would be desirable to fabricate planar lightwave circuits (PLC) that combine "conventional" dry etched structures and crystal-aligned structures obtained by preferential etching. In this section, we present a self-aligning fabrication process that allows for the realization of such combined PLC.

In a first step, the PLC is subdivided into a crystal-aligned section and an arbitrarily oriented section. Within the crystal-aligned section, all sidewalls coincide with stable planes of the crystal, and no convex corners occur. This domain is well suited for waveguides which require smooth sidewalls and a precise geometry, e.g., long straight waveguide of low-loss delay lines, slot or strip waveguides for which strong interaction with the cover material is desired, or PhC sections with special cavities as discussed in the last section. The arbitrarily oriented sections may contain structures with contours of arbitrary orientation, that may be curved or have convex corners. Such structures can, e.g., comprise waveguide bends, MMI couplers or tapers.

An example of a combined planar lightwave circuit (PLC) for a $\langle 110 \rangle$-oriented device layer is depicted in Fig. 5.9. The planar lightwave circuit 30 comprises crystal-aligned sections 11a, 11b, 11c in which the plane sidewalls of the strip waveguides 42, 44, 52, 54

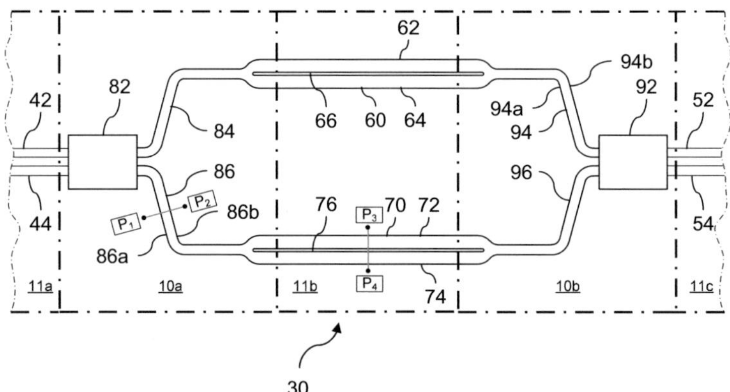

Fig. 5.9. Example of crystal-oriented devices embedded into conventional feed waveguide structures: The planar lightwave circuit 30 comprises domains 11a, 11b, 11c in which the plane sidewalls of the strip waveguides 42, 44, 52, 54 and the slot waveguides 60, 70 are aligned with the crystal lattice, and domains 10a, 10b with curved sidewalls 86a, 86b, 94a, 94b of arbitrary orientation and multi-mode interference (MMI) couplers 82, 92 with convex corners. To avoid offsets between the domains 10, 10b and 11a, 11b, 11c, a self-aligning fabrication technique is used. The processing steps are depicted in Fig. 5.10 by sketching the cross sections along the lines P_1P_2 and P_3P_4.

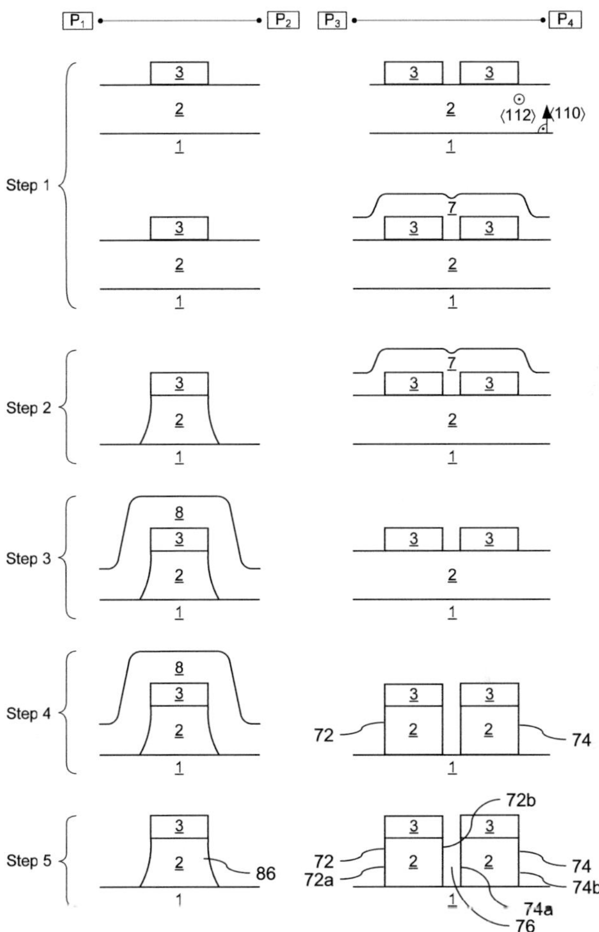

Fig. 5.10. Processing steps of the self-aligning fabrication technique. The cross sections along the lines P_1P_2 and P_3P_4 are defined in Fig. 5.9. (1) Silicon dioxide buffer layer; (2) SOI device layer; (3) Primary etch mask for waveguide structures; (7) Secondary etch mask to protect the crystal-aligned parts of the structure during anisotropic (dry) etching; (8) Secondary etch mask to protect the arbitrarily oriented parts of the structure during preferential etching; in the crystal-aligned domain (right), the sidewalls 72a, 72b, 74a, 74b of the strips 72, 74 and of the slot 76 are defined by {111} planes of the silicon lattice.

and the slot waveguides 60, 70 are aligned with the crystal lattice. The arbitrarily oriented sections 10a, 10b comprise curved sidewalls 86a, 86b, 94a, 94b and MMI couplers 82, 92.

A crucial issue is the transition between the crystal-aligned sections and the arbitrarily oriented ones. To avoid offsets, the primary etch mask for both domains has to be structured in one go. In subsequent processing steps, either the crystal-aligned or the arbitrarily oriented domain are temporarily protected by secondary etch masks. The processing steps for the PLC shown in Fig. 5.9 are illustrated in Fig. 5.10 by sketching the cross sections along the lines P_1P_2 and P_3P_4. In a first step, the primary mask (3) is deposited and structured, thereby defining the waveguides both in the crystal-aligned and in the arbitrarily oriented domains. The crystal-aligned domain (right) is then covered by a secondary protection mask (7). Anisotropic dry etching is used to structure the arbitrarily oriented domain, thereby creating sidewalls which might not be perfectly vertical, step 2. in step 3, the mask (7) is removed and another secondary mask (8) is deposited on the arbitrarily-oriented domain (left). Preferential wet etching is used to structure the crystal-aligned domain, resulting in perfectly smooth and vertical {111} sidewalls 72a, 72b, 74a, 74b of the strips 72, 74 and of the slot 76, see step 4. Finally the secondary mask (8) is removed by selective etching.

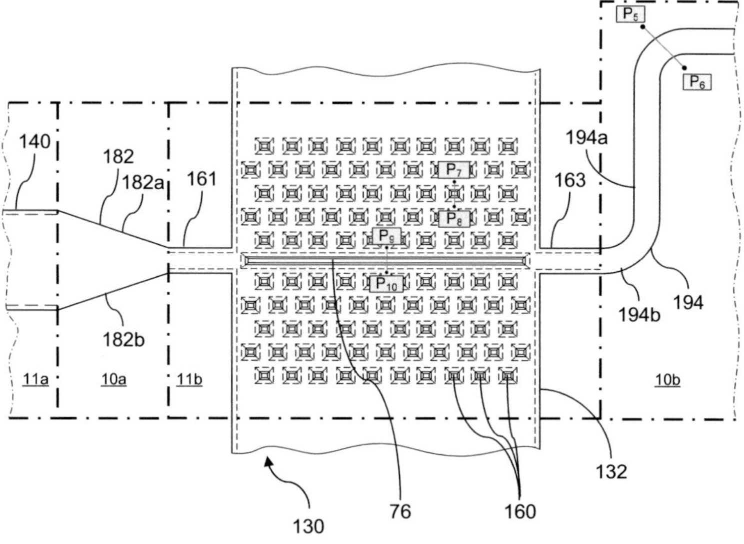

Fig. 5.11. Example of crystal-oriented devices embedded into conventional feed waveguide structures: The planar lightwave circuit 130 comprises a wide strip waveguide 140, a taper 182, narrow strip waveguides 161, 163, a photonic crystal (PhC) section 132 and curved waveguide sections 194. Within the photonic crystal structure, light is guided along a defect waveguide which comprises a slot 76. In the domains 11a, 11b, the plane sidewalls of the strip waveguides 140, 161, 163, the sidewalls of the PhC cavities 160 and of the slot 76 are aligned with the crystal lattice. In the domains 10a, 10b, the sidewalls 1821, 182b of the taper 182 and the sidewalls 194a, 194b of the curved section 194 are arbitrarily orientated and/or curved. The processing steps are depicted in Fig. 5.12 by sketching the cross sections along the lines P_5P_6, P_7P_8 and P_9P_{10}.

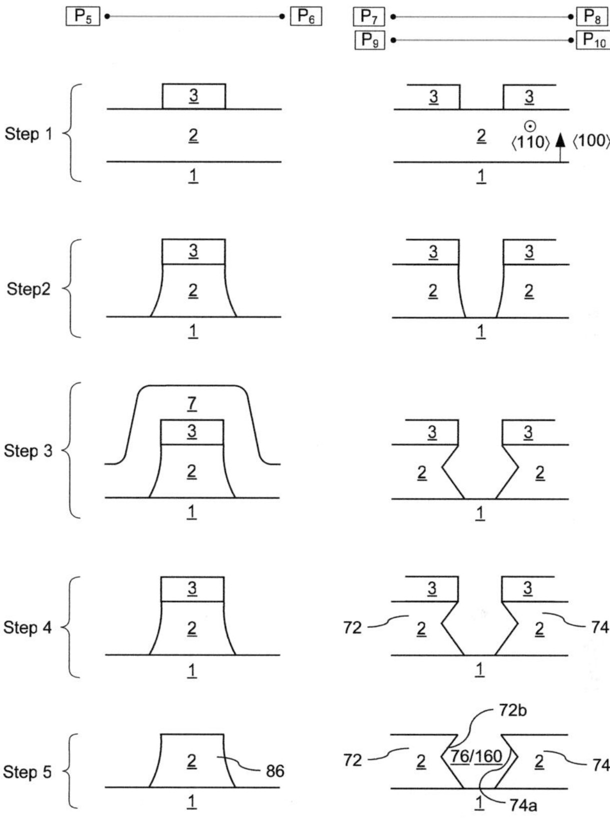

Fig. 5.12. Processing steps of the self-aligning fabrication technique. The lines P_5P_6, P_7P_8 and P_9P_{10} are defined in Fig. 5.11. (1) Silicon dioxide buffer layer; (2) SOI device layer; (3) Primary etch mask for waveguide structures; (7) Secondary etch mask to protect arbitrarily oriented parts of the structure during preferential etching; the sidewalls 72a, 72b of the slot 76 and the cavities 160 are defined by ⟨111⟩-planes of the silicon crystal lattice.

An example of a combined PLC for a $\langle 100 \rangle$-oriented device layer is depicted in Fig. 5.11. The PLC 130 comprises again crystal-aligned sections 11a, 11b and arbitrarily oriented sections 10a, 10b. Section 11b contains a PhC structure 132, consisting of a regular arrangement of cavities 160 as depicted in Fig. 5.4. The arbitrarily oriented sections 10a and 10b contain a waveguide taper 182 and waveguide bends 194. The steps of the self-aligning fabrication process are depicted in Fig. 5.12 by sketching the cross sections along the lines P_5P_6, P_7P_8 and P_9P_{10}. In contrast to the process depicted in Fig. 5.10, the crystal-aligned domains (right) are not protected during the dry etching process (step 2), resulting in oktaeder-like cavities and slots with rhombic cross sections, see also Figs 5.2 and 5.4.

5.5   Fabrication of Waveguides

In this section, we report on the fabrication of prototype waveguide structures by using KOH-based preferential wet etching of silicon. We show scanning electron microscope (SEM) pictures that prove that the proposed fabrication processes are well suited for the fabrication of nanometer-scale optical waveguides with complex geometries.

5.5.1   Process Parameters

Straight waveguides in a $340\,\mathrm{nm}$ thick $\langle 100 \rangle$-oriented SOI device layer were fabricated by a single preferential wet etch step (process 1), and by a combination of dry etching and preferential wet etching (process 2). In this section we give technological details for each of the two processes. The results are presented in the next section.

In both cases, only small amounts of material are removed, and excessive formation of H_2 bubbles on the etched surface was not observed. It was hence not necessary to apply ultrasound during the etch. In some cases, ultrasound treatment even turned out to enhance underetching and to destroy the Si_3N_4 mask.

Process 1: Single Preferential Wet Etch Step

As a first step a $90\,\mathrm{nm}$ thick layer of stoichiometric Si_3N_4 is first deposited by low-pressure chemical vapor deposition (LPCVD). The samples ($10 \times 10\,\mathrm{mm}^2$ chips) are then treated with an adhesion promoter (Hexamethyldisilazane, HMDS) to improve resist adhesion. A $250\,\mathrm{nm}$ thick layer of negative-tone electron beam resist (ma-N2403, Microresist, http://www.microresist.de) is spin-coated and dried on a hot plate. The resist is exposed by direct-write electron-beam lithography performed on a Raith100 system at $20\,\mathrm{kV}$ acceleration voltage. Depending on the width of the waveguide strips, ideal doses range from $100\,\mu\mathrm{C/cm}^2$ to $150\,\mu\mathrm{C/cm}^2$. The resist structures are developed in a TMAH-based developer (maD-532, Microresist, http://www.microresist.de). Thorough rinsing ($> 5\,\mathrm{min}$) with deionized water turned out to be crucial for good contrast of the developed structures. The resist is hard-baked and then treated with an O_2 plasma to reduce resist roughness and to strip undeveloped residues. The Si_3N_4 mask is then opened in an reactive ion etching process based on CF_4, CHF_3, and O_2 (STS Multiplex), and the resist is stripped in an O_2 plasma. To remove the native oxide on the surface of the silicon device layer, the chips are dipped in 20:1 H_2O:HF(40%) for 20 s. Preferential wet etching is finally performed at different temperatures in aqueous solutions containing 45 % KOH by mass. The Si_3N_4 mask is removed by selective wet etching in hot concentrated phosphoric acid (H_3PO_4, 85 %).

Process 2: Combination of Dry Etching and Preferential Wet Etching

For waveguide structures that require prestructuring by anisotropic dry etching, a two-step process was developed. The lithography and the structuring of the hard mask definition were performed in cooperation with the Fraunhofer-Institut für Nachrichtentechnik, Heinrich-Hertz-Institut in Berlin. The dry etching was done at the Technical University Berlin. Process 2 comprises the following steps: A 70 nm thick hard mask of Si_3N_4 is first deposited on a 100 mm wafer deploying a plasma-enhanced chemical vapor deposition process that has been optimized to obtain pinhole-free layers. The mask layer is then coated with negative-tone electron beam resist (ma-N2405, Microresist, http://www.microresist.de) and exposed with 50 kV direct write electron beam lithography. Hard-bake and descum steps follow, and the SiN_x hard mask is structured by RIE. In a second RIE process, the silicon device layer is etched , and the resist then stripped in an O_2 plasma. The wafer is then diced into smaller chips, and native oxide is removed in an HF dip. Preferential wet etching is finally performed in aqueous solutions containing 45 % KOH by mass. The hard mask is removed by selective wet etching in hot concentrated phosphoric acid (H_3PO_4, 85 %).

5.5.2   Prototype Waveguides

Scanning electron microscope (SEM) pictures of the fabricated waveguides are shown in Figs. 5.13 and 5.14. The white dirt particles seen in the SEM pictures are presumably due to etch chemicals of insufficient purity grade. Unfortunately, this made it impossible to measure the optical loss of the prototype waveguides. Using VLSI or ULSI grade KOH (rather than analytical grade) and working under strict cleanroom conditions will solve this problem.

The trapezoidal waveguides shown in Fig. 5.13 were fabricated in a single preferential wet etch step as described in Process 1 at 60 °C, but without removing the Si_3N_4 mask. These waveguides were used to investigate the effects of misalignment between the edges of the etch mask and the corresponding crystal direction. The complete structure consisted of 121 straight waveguides in a fan-like arrangement. The inclinations with respect to the $\langle 110 \rangle$ direction cover the range between $-6°$ and $+6°$ in steps of 0.1 ° For a mask width of 150 nm, a misalignment of more than 5 ° leads to severe underetching and the mask is destroyed, see Fig. 5.13 (a). For a misalignment of approx. 2.5 °, the sidewalls feature plateaus of stable {111} planes that are separated by steps, see Fig. 5.13 (b). For ±0.1 ° of misalignment, however, the sidewalls are perfectly smooth, and no roughness is visible in the scanning electron microscope. Considerable underetching can be observed from the end of the waveguide. This demonstrates the instability of any convex corners exposed to the etchant.

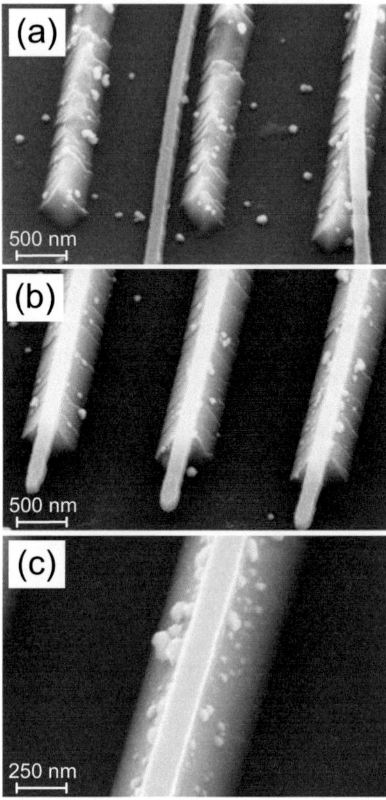

Fig. 5.13. SEM pictures of trapezoidal SOI waveguides in a $\langle 100 \rangle$-oriented device layer (top width $w_{\text{top}} = 150$ nm); The structures were etched at 60 in an aqueous solution that contains 45 % KOH by mass. (a) Waveguides strongly misaligned by $-5.6°$, $-5.5°$, and $-5.4°$ (from left to right) with respect to the $\langle 110 \rangle$-direction; (b) Waveguides misaligned by $-2.6°$, $-2.5°$, and $-2.4°$ (from left to right) with respect to the $\langle 110 \rangle$-direction; (c) Waveguide well aligned with the $\langle 110 \rangle$-direction ($\pm 0.1°$ deviation)

Figure 5.14 (a) shows a prototype waveguide with a cross section according to Fig. 5.2 (b). The structure has been fabricated according to Process 1 at 25 °C, and the Si_3N_4 mask was left in place. Apart from dirt particles, the sidewalls are again perfectly smooth, even though the etch mask is rough. The quality of sidewalls obtained by preferential wet etching can hence exceed the quality of the etch mask, which is always imperfect due to limited lithographic resolution. The concave edge at the bottom of the groove in Fig. 5.14 looks perfectly sharp.

Fig. 5.14. SEM pictures of SOI waveguides with special cross sections in a $\langle 100 \rangle$-oriented device layer, see Fig. 5.2 (b).

The cross section of the slot waveguide depicted in Fig. 5.14 (b) is sketched in Fig. 5.2 (d). The waveguide was fabricated according to Process 2 with a temperature of the wet etchant of 25 °C. The SiN_x mask was removed afterwards. Again, all sidewalls look again atomically smooth, and concave edges are perfectly sharp. Diffusion-induced reduction of etch rates in high-aspect ratio slots could not be observed.

5.6   Summary

We propose a novel fabrication method for creating highly precise waveguides with special cross-sectional geometries and smooth sidewalls. The fabrication is based on preferential etching of crystalline materials, whereby the sidewalls of the waveguide structure coincide with stable planes of the crystal. The chemistry of preferential wet etching is discussed for semiconductor materials, with an emphasis on silicon, and suitable process parameters are identified. We present a variety of different waveguide geometries that can be fabricated by preferential etching of crystalline SOI device layers. We propose a self-aligning fabrication process that allows to combine preferential etching techniques with conventional anisotropic dry etching. Prototype structures were produced and investigated by scanning electron beam microscopy, and the effects of misalignment between the structure and the crystallographic directions are discussed. In our experiments, the quality of the sidewalls exceeds the quality of the etch mask. The results indicate that optical waveguides even with complex cross sectional geometries can be fabricated with nanometer-scale accuracy by using preferential wet etching techniques.

Chapter 6

Second-Order Nonlinear Devices: Fast SOI-Based Electro-Optic Modulators

6.1 Introduction

Optical 100 Gbit/s-Ethernet requires a new generation of low-cost electro-optic modulators that can be mass-produced and integrated into silicon-based electronic microprocessor circuitry. These devices should exhibit low power consumption and bandwidths of several tens of Gigahertz.

So far, the fastest modulation in silicon-on-insulator (SOI) waveguides was achieved by free-carrier injection. Modulation at 20 GHz in a silicon-based device was recently demonstrated with a reverse-biased pn-junction, allowing data transmission up to 30 Gbit/s [69]. The size and power consumption of such devices can be considerably reduced by exploiting slow light in photonic crystal (PhC) waveguides [42]. However, the speed is inherently limited by the dynamics of the free carriers.

On the other hand, bandwidths of 110 GHz have been demonstrated in modulators based on the linear electro-optic effect (Pockels effect) in polymers [18]. Electro-optic coefficients of such organic materials can exceed 500 pm/V [20]. A first attempt to deploy electro-optic polymers on SOI has been published [8], but the bandwidth is still limited to the low MHz-regime due to large RC time constants.

In this chapter, we propose novel schemes for compact SOI-based modulators with large electrical and optical bandwidth and low power consumption. We exploit electro-optic interaction in a cover material that is deposited onto a pre-structured silicon waveguide. Our calculations show that these devices can work at and above bitrates of 100 Gbit/s.

6.2 Novel SOI Modulator Schemes

Figures 1 and 2 show schematic views of the three novel modulator schemes. In Fig. 6.1, the light is guided by silicon strips which are embedded in the electro-optic material (EO).

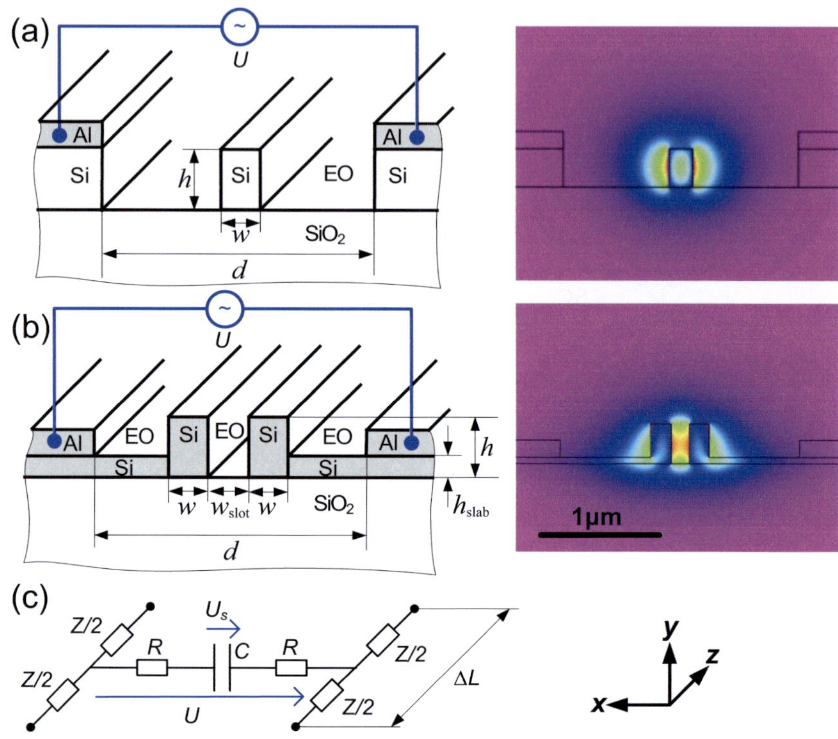

Fig. 6.1. Novel waveguide-based SOI modulator schemes. The optical slot waveguides consist of silicon ribs on top of the buried oxide (SiO_2) and are covered with an electro-optic material EO. Conductive cross-sectional domains (containing aluminum or doped silicon) are indicated by shading. For both structures, the $\mathcal{E}_x$-component of the optical mode fields are depicted on the right. (a) Strip waveguide structure, $w = 200\,\text{nm}$, $h = 340\,\text{nm}$, $d = 2\,\mu\text{m}$; (b) Slot waveguide structure, $w = 170\,\text{nm}$, $w_{\text{slot}} = 150\,\text{nm}$, $h = 340\,\text{nm}$, $h_{\text{slab}} = 50\,\text{nm}$, $d = 2\,\mu\text{m}$; (c) Lumped-element model a short section (length ΔL) of the slot waveguide structure

Field discontinuities at the high index-contrast interfaces lead to field enhancements that provide strong interaction of the guided mode with the electro-optic material, see field plots in Fig. 6.1.

In the scheme of Fig. 6.1 (a) the microwave field is applied via two aluminum conductor paths running in parallel to the optical strip waveguide. The spacing is chosen large enough (typically $\geq 1\,\mu\text{m}$) to avoid optical loss. For the slot waveguide, Fig. 6.1 (b), both silicon strips are doped and connected to the aluminum conductors by thin silicon slabs. Arsenic doping with a density of $n_D \approx 2 \times 10^{16}\,\text{cm}^{-3}$ yields sufficient electrical conductivity $\sigma_{\text{Si}} \approx 10\,(\Omega\,\text{cm})^{-1}$ [78], but does not induce relevant optical loss [99].

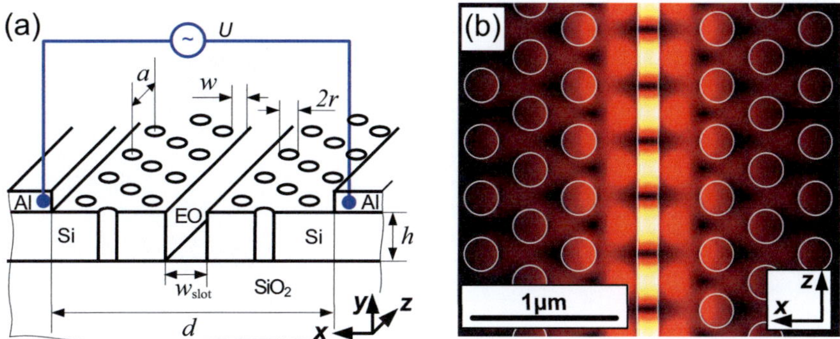

Fig. 6.2. (a) PhC-based modulator scheme: A PhC line defect waveguide with a slot is etched into the SOI device layer. The structure is covered with an electro-optic material EO. The conductive PhC regions consist of doped silicon and are electrically connected to the aluminum conductor paths. (b) Corresponding field plot (E_x); $a = 380\,\mathrm{nm}$, $h = 340\,\mathrm{nm}$, $2r = 198\,\mathrm{nm}\ldots212\,\mathrm{nm}$, $w = 263\,\mathrm{nm}$, $w_{\mathrm{slot}} = 150\,\mathrm{nm}$, $d = 2\,\mu\mathrm{m}$;

Figure 6.1 (c) depicts a lumped element model of a short segment (length ΔL) of the slot waveguide configuration: $Z \propto \Delta L$ denotes the series impedance for each of the conductor tracks, whereas $R \propto 1/\Delta L$ and $C \propto \Delta L$ define the shunt admittance of the silicon slabs and ribs. For a DC signal, the applied voltage $U = U_s$ drops over the narrow slot (typical width: 150 nm), where high field strengths can be easily achieved.

The optically isolating slab regions in Fig. 6.1 (b) can be replaced by a photonic crystal (PhC) structure. This scheme is depicted in Fig. 6.2 (a). The optical intensity is high in the electro-optic material EO that fills the slot, see Fig. 6.2 (b). As for the slab structure, the modulation frequency is limited by the RC time constant. The group velocity of the optical signal can be significantly reduced by an appropriate design of the PhC. The interaction time with the electro-optic material can thus be increased; this decreases the operating voltage and/or the device length.

6.3 Operating Voltage

For the case of an appropriately oriented second-order nonlinear material EO, e.g., a polymer, a microwave electric field induces a local change of the dielectric profile seen by the optical field. For electro-optic polymers, r_{33} is normally much larger than all the other electro-optic coefficients [20], i. e., the axis of strongest electro-optic interaction corresponds to w in Eq. (1.76). In the following, we assume that the electro-optic material has isotropic optical properties if no electric field is applied. The dielectric tensor $\underline{\varepsilon}_{r,0}$ in Eq. (1.75) can then be replaced by a scalar $\epsilon_0 n_{\mathrm{EO}}^2$, where n_{EO} denotes the refractive index of the electro-optic polymer. We further assume that the electro-optic polymer is poled such that the axis of strongest electro-optic interaction is aligned along the x-direction.

Neglecting all electro-optic coefficients except for r_{33}, the change of the permeability tensor can be written with respect to the coordinate system (x, y, z),

$$\Delta\underline{\epsilon}_r(\mathbf{r}, t) = -\epsilon_0 n_{\mathrm{EO}}^4 r_{33} \mathbf{E}_{\mathrm{mw},x}(\mathbf{r}, t) \begin{pmatrix} 1 & 0 & 0 \\ 0 & 0 & 0 \\ 0 & 0 & 0 \end{pmatrix}. \tag{6.1}$$

$\mathbf{E}_{\mathrm{mw},x}(\mathbf{r}, t)$ denotes the x-component of the microwave field $\mathbf{E}_{\mathrm{mw}}(\mathbf{r}, t)$ travelling along the conductors. In this representation, the microwave signal has zero carrier frequency $\omega_0 = 0$, and the time dependent modulation is represented by a voltage signal $u(z, t)$ travelling along the z-direction. For the moment, the microwave mode field $\mathcal{E}(x, y, \omega_0)$ is assumed to be independent of electric modulation frequency. $\mathbf{E}_{\mathrm{mw}}(\mathbf{r}, t)$ can then be represented by

$$\mathbf{E}_{\mathrm{mw}}(\mathbf{r}, t) = \frac{u(z, t)}{\mathcal{U}} \mathcal{E}(x, y, \omega_0). \tag{6.2}$$

The voltage amplitude $\mathcal{U}$ is the voltage that is associated with the mode field $\mathcal{E}(x, y, \omega_0)$. According to Eq. (1.68), the change $\Delta\beta(z, t)$ of the propagation constant can now be written as

$$\Delta\beta(z, t) = -\frac{u(z, t)}{\mathcal{U}} \frac{n_{\mathrm{EO}} k_0}{2 P(\omega_c) Z_0} \iint \frac{1}{2} n_{\mathrm{EO}}^3 r_{33} \mathcal{E}_x(x, y, \omega_0) |\mathcal{E}_x(x, y, \omega_c)|^2 \, \mathrm{d}x\, \mathrm{d}y. \tag{6.3}$$

The quantity $k_0 = \omega_c/c$ denotes the free-space wave number of the optical signal with carrier frequency ω_c. $\mathcal{E}_x(x, y, \omega_0)$ denotes the x-component of the microwave mode field, and $\mathcal{E}_x(x, y, \omega_c)$ corresponds to the x-component of the optical mode field. $\mathcal{U}$ is the electrical potential difference between the conductor paths that is associated with the microwave mode field $\mathcal{E}(x, y, \omega_0)$.

For an electrical DC voltage u_{DC} applied to the conductor paths, $u(z, t) = u_{\mathrm{DC}}$ is constant along the waveguide. In this case, the accumulated optical phase shift $\Phi = \Delta\beta L$ is proportional to the length L of the waveguide and to the applied voltage u_{DC}. The normalized phase shift $\Pi^\times = \Phi/(u_{\mathrm{DC}} L)$ can thus be written as

$$\Pi^\times = \frac{1}{\mathcal{U}} \frac{n_{\mathrm{EO}}^4 k_0 r_{33}}{4 P(\omega_c) Z_0} \iint \mathcal{E}_x(x, y, \omega_0) |\mathcal{E}_x(x, y, \omega_c)|^2 \, \mathrm{d}x\, \mathrm{d}y \tag{6.4}$$

This result is very intuitive: The x-component $\mathcal{E}_x(x, y, \omega_0)$ of the microwave field modulates the refractive index seen by the x-component $\mathcal{E}_x(x, y, \omega_c)$ of the optical field. The electro-optic interaction hence increases with the spatial overlap of the optical and the electrical field. This is expressed by the overlap integral on the right-hand side of Eq. (6.4).

Table 6.1 lists characteristic data for the three different device configurations. In the first column, the group velocity v_{g} for $\lambda = 1.55\,\mu\mathrm{m}$ is specified in fractions of the vacuum speed of light c. Assuming a polymer with a moderate $r_{33} = 50\,\mathrm{pm}\,/\,\mathrm{V}$ [20], the normalized phase shift $\Pi^\times$ has been calculated for the strip and the slot waveguide, and for the PhC structure, see second column of Table 6.1. As expected, the slot waveguide structure has a higher normalized phase shift than the simple strip due to more closely spaced electrodes. For the PhC modulator, the normalized phase shift is further increased by the low optical group velocity. For a Mach-Zehnder modulator in push-pull configuration, a phase shift

Configuration	v_g/c	$\Pi^\times$ $\left[(\mathrm{V\,mm})^{-1}\right]$	$L_{\pi/4}$ [mm]	f_{gd} [GHz]	f_{RC} [GHz]
Strip	0.392	~ 0.14	7.5	–	–
Slot	0.377	~ 0.86	1.2	–	**110**
PhC	0.054	4.44	0.26	**40**	280

Table 6.1. Characteristic data of the modulator based on a strip waveguide, a slot waveguide, and a slotted PhC line defect waveguide (wavelength $\lambda = 1.55\,\mu\mathrm{m}$, electro-optic coefficient $r_{33} = 50\,\mathrm{pm/V}$). The lengths $L_{\pi/4}$ refer to a phase shift of $\Phi = \pi/4$ and to a peak-to-peak voltage of $U = 1.5\,\mathrm{V}$. The group delay-related bandwidth f_{gd} increases with U.

amplitude of $\pi/4$ is needed in each arm for complete extinction. Assuming a peak-to-peak voltage of 1.5 V, the device lengths $L_{\pi/4}$ for a $\pi/4$ phase shift have been calculated, see third column of Table 6.1. We find that devices can be as short as $260\,\mu\mathrm{m}$ (1.2 mm) for the PhC (slot waveguide) configuration. In all cases, the refractive index change due to free carriers is much weaker than the electro-optic effect and can therefore be neglected.

6.4 Bandwidth Limitations

An electro-optic amplitude modulator is obtained by incorporating the above-mentioned electro-optic phase modulators into an interferometric configuration, e. g., a Mach-Zehnder interferometer (MZI). To allow for push-pull operation, the upper and the lower arm of the MZI comprise each a phase modulator, denoted as $\mathrm{PM_u}$ and $\mathrm{PM_l}$, respectively, see Fig. 6.3. The optical phase shifts imposed in the upper and lower arm are denoted as $\Phi_\mathrm{u}(t)$ and $\Phi_\mathrm{l}(t)$, respectively. The constant phase shift Φ_op in the upper arm defines the operation point of the amplitude modulator. A constant optical input power of P_in is fed into the MZI, and the modulated output power $P_\mathrm{out}(t)$ is detected by measuring the

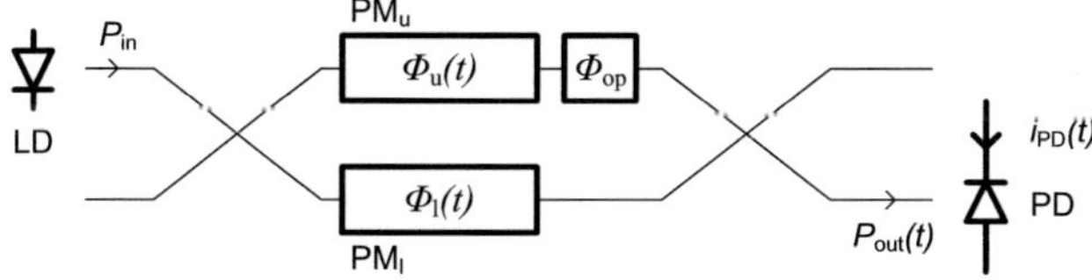

Fig. 6.3. Mach-Zehnder amplitude modulator. The laser diode LD launches a constant optical input power of P_in to Mach-Zehnder interferometer. Optical phase shifts $\Phi_\mathrm{u}(t)$ and $\Phi_\mathrm{l}(t)$ are imposed in the upper and lower arm by the phase modulators $\mathrm{PM_u}$ and $\mathrm{PM_l}$, respectively. The constant phase shift Φ_op in the upper arm defines the operation point of the modulator. The modulated output power $P_\mathrm{out}(t)$ is detected by measuring the photocurrent $i_\mathrm{PD}(t)$ at the output of a photodetector PD.

photocurrent $i_{\mathrm{PD}}(t) = SP_{\mathrm{out}}(t)$ at the output of a photodetector[1] PD with responsivity S. Neglecting optical propagation loss in the modulator, $P_{\mathrm{out}}(t)$ is given by

$$P_{\mathrm{out}}(t) = P_{\mathrm{in}} \cos^2\left(\frac{\Phi_{\mathrm{u}}(t) - \Phi_{\mathrm{l}}(t) + \Phi_{\mathrm{op}}}{2}\right). \tag{6.5}$$

For push-pull operation, $\Phi_{\mathrm{op}} = \pi/2$ is chosen and the phase modulators $\mathrm{PS_u}$ and $\mathrm{PS_l}$ are operated in differential mode,

$$\Phi_{\mathrm{u}}(t) = -\Phi_{\mathrm{l}}(t) =: \Phi(t). \tag{6.6}$$

The modulated optical power can then be written as

$$P_{\mathrm{out}}(t) = \frac{P_{\mathrm{in}}}{2}\left(1 + \sin\left(2\Phi(t)\right)\right), \tag{6.7}$$

In the following, we only consider the phase modulator $\mathrm{PM_u}$ in the upper arm. The relations for the phase modulator $\mathrm{PM_u}$ in the lower arm can be easily obtained by swapping the signs for the phase and the voltage.

To measure the bandwidth of the modulator, a sinusoidal electric signal $u(t)$ is applied to the phase modulator,

$$u(t) = \widehat{U}\cos\left(\Omega t\right). \tag{6.8}$$

Neglecting any propagation delay, the corresponding optical phase shift $\Phi(t)$ can then be written as

$$\Phi(t) = \widehat{\Phi}(\Omega)\cos\left(\Omega t\right), \tag{6.9}$$

where the phase modulation amplitude $\widehat{\Phi}(\Omega)$ represents the frequency characteristic of the phase modulator. The resulting photocurrent reads

$$i_{\mathrm{PD}}(t) = \frac{SP_{\mathrm{in}}}{2}\left(1 + \sin\left(2\widehat{\Phi}(\Omega)\cos\left(\Omega t\right)\right)\right). \tag{6.10}$$

The current $i_{\mathrm{PD}}(t)$ is periodic in time with period $2\pi/\Omega$. Using Eq. (E.9), it can be expanded in a Fourier series,

$$i_{\mathrm{PD}}(t) = I_0\left(\Omega\right) + \sum_{\nu=1}^{\infty} I_\nu\left(\Omega\right)\sin\left(\nu\left(\Omega t - \frac{\pi}{2}\right)\right), \tag{6.11}$$

where the Fourier coefficients $I_\nu\left(\Omega\right)$ are given by

$$I_\nu\left(\Omega\right) = \begin{cases} SP_{\mathrm{in}}/2 & \text{for} \quad \nu = 0 \\ SP_{\mathrm{in}}J_\nu\left(2\widehat{\Phi}(\Omega)\right) & \text{for} \quad \nu \text{ odd} \\ 0 & \text{for} \quad \nu \text{ even}, \nu \neq 0 \end{cases}. \tag{6.12}$$

J_ν denotes the νth order Bessel function of the first kind.

---

[1] For the analysis, an ideal detector of infinite bandwidth is assumed. If the frequency characteristic of the detector is measured in a real experiment, the characteristic of the photodetector must be eliminated by calibration.

The amplitude $\widehat{U}$ is chosen such that complete extinction is obtained for low frequencies, i. e., $\lim_{\Omega\to 0} \widehat{\Phi}(\Omega) = \widehat{\Phi}(0) = \pi/4$. For the fundamental modulation frequency component I_1 of the photocurrent we thus find the low-frequency limit

$$\lim_{\Omega\to 0} I_1(\Omega) = \frac{SP_{\mathrm{in}}J_1(\pi/2)}{2}. \tag{6.13}$$

If the modulation frequency Ω increases, the phase shift $\widehat{\Phi}(\Omega)$ degrades, which translates into a decrease of the fundamental frequency component $I_1(\Omega)$. At the 3 dB-point we have

$$\frac{J_1\left(2\widehat{\Phi}(\Omega_{3\,\mathrm{dB}})\right)}{J_1\left(2\widehat{\Phi}(0)\right)} = \frac{1}{\sqrt{2}}, \qquad \widehat{\Phi}(0) = 0.2500\pi \quad \Rightarrow \quad \widehat{\Phi}(\Omega_{3\,\mathrm{dB}}) = 0.1410\pi. \tag{6.14}$$

The 3 dB-bandwidth $f_{3\,\mathrm{dB}} = \Omega_{3\,\mathrm{dB}}/(2\pi)$ of the modulator is thus defined as the frequency for which the phase shift $\widehat{\Phi}$ has degraded from its low-frequency limit $\pi/4$ to the value 0.1410π. For the photocurrent, this corresponds to a reduction of the fundamental modulation frequency component by a factor of $1/\sqrt{2}$ (3 dB).

6.4.1 RC Time Constants

If the slot waveguide or the PhC modulator are fed with an AC signal, the capacity of the slot has to be charged via the resistive slab regions, $U_{\mathrm{s}} < U$. This is apparent from the lumped element model of a short slot waveguide segment, Fig. 6.1 (c). The phase shift is proportional to U_{s}. The associated RC-limited 3 dB-bandwidth according to Eq. (6.14) is hence defined by $U_{\mathrm{s}}/U = 0.1410\pi/0.2500\pi$ and amounts to

$$f_{RC} = \frac{0.1166}{RC}. \tag{6.15}$$

R and C define the shunt admittance of the silicon slabs and ribs.

6.4.2 Velocity Mismatch and Travelling-Wave Configuration

Apart from RC time constants, the modulation bandwidth of the devices is limited if the group velocity $v_{\mathrm{g,el}}$ of the electrical signal and the group velocity $v_{\mathrm{g,op}}$ of the optical signal differ: The corresponding spatial walk-off leads to a degradation of the modulator performance for high frequencies. This effect shall now be investigated in more detail.

The microwave field is represented as in Eq. (6.2), and for the optical field we use the usual ansatz

$$\underline{\mathbf{E}}_{\mathrm{opt}}(\mathbf{r}, t) = A(z, t, \omega_{\mathrm{c}})\frac{\mathcal{E}(x, y, \omega_{\mathrm{c}})}{\sqrt{\mathcal{P}(\omega_{\mathrm{c}})}}\, e^{\mathrm{j}(\omega_{\mathrm{c}}t - \beta(\omega_{\mathrm{c}})z)}. \tag{6.16}$$

Neglecting optical rectification, the optical field does not have any influence on the microwave field, and Eq. (1.49) can be adopted to describe the propagation of the microwave mode. We neglect dispersion of the microwave signal and obtain

$$\frac{\partial u(z, t)}{\partial z} + \beta_{\mathrm{mw}}^{(1)}\frac{\partial u(z, t)}{\partial t} = 0, \tag{6.17}$$

The quantity $\beta_{\text{mw}}^{(1)} = \frac{\mathrm{d}\beta}{\mathrm{d}\omega}\big|_{\omega=0}$ is the inverse of the microwave group velocity. The optical signal is influenced by the microwave field via the space and time-dependent change $\Delta\beta(z,t)$ of the propagation constant,

$$\frac{\partial A(z,t)}{\partial z} + \beta_{\text{opt}}^{(1)}\frac{\partial A(z,t)}{\partial t} = -\,\mathrm{j}\,\Delta\beta(z,t)A(z,t), \tag{6.18}$$

where $\beta_{\text{opt}}^{(1)} = \mathrm{d}\beta/\mathrm{d}\omega\big|_{\omega=\omega_c}$ is the inverse of the optical group velocity. The differential equations (6.17) and (6.18) are coupled via the fact, that $\Delta\beta(z,t)$ depends on $u(z,t)$, see Eq. (6.3).

We now introduce a retarded time frame that travels along with the group velocity of the optical signal, see Eq. (1.54)

$$t' = t - \beta_{\text{opt}}^{(1)}z. \tag{6.19}$$

All quantities that refer to this retarded time frame are marked with a prime in the following.

For the microwave field. Eq. (6.17) transforms to

$$\frac{\partial u'(z,t')}{\partial z} + \left(\beta_{\text{mw}}^{(1)} - \beta_{\text{opt}}^{(1)}\right)\frac{\partial u'(z,t')}{\partial t'} = 0 \tag{6.20}$$

and is solved by

$$u'(z,t') = u'(0, t' - \Delta\beta^{(1)}z), \qquad \Delta\beta^{(1)} = \beta_{\text{mw}}^{(1)} - \beta_{\text{opt}}^{(1)} \tag{6.21}$$

The quantity $\Delta\beta^{(1)}z$ can be interpreted as the group delay difference $\Delta\tau_g$ that the microwave signal and the optical signal acquire over a propagation distance z.

For the optical field, Eq. (6.18) reduces to

$$\frac{\partial A'(z,t')}{\partial z} = -\,\mathrm{j}\,\Delta\beta'(z,t')A'(z,t'), \tag{6.22}$$

and can be directly integrated,

$$A'(z,t') = A'(0,t')\,\mathrm{e}^{\mathrm{j}\,\Phi'(z,t')}, \qquad \Phi'(z,t') = -\int_0^z \Delta\beta'(\zeta,t')\,\mathrm{d}\zeta. \tag{6.23}$$

The change of the propagation constant $\Delta\beta'(z,t')$ is obtained by inserting Eq. (6.21) into Eq. (6.3). The accumulated phase shift $\Phi'(z,t')$ for the optical signal thus amounts to

$$\Phi'(z,t') = -\Pi^\times \int_0^z u'(0, t' - \Delta\beta^{(1)}\zeta)\,\mathrm{d}\zeta, \tag{6.24}$$

where the normalized phase shift $\Pi^\times$ is defined by Eq. (6.4).

To study the degradation of the phase shift with increasing microwave modulation frequency Ω, we investigate a harmonic signal of the form

$$u'(0,t') = \widehat{U}\cos\left(\Omega t'\right). \tag{6.25}$$

For a waveguide of length L, the accumulated phase shift $\Phi'(L, t')$ is obtained by inserting Eq. (6.25) into Eq. (6.24),

$$\Phi'(L, t') = \widehat{\Phi}(L, \Omega) \cos\left(\Omega\left(t' - \frac{\Delta\beta^{(1)}L}{2}\right)\right). \tag{6.26}$$

The amplitude $\widehat{\Phi}(L, \Omega)$ of the phase modulation depends on the modulation frequency Ω and on the waveguide length L via

$$\widehat{\Phi}(L, \Omega) = -\Pi^{\times}\,\widehat{U}\,L\,\mathrm{sinc}\left(\frac{\Omega\Delta\beta^{(1)}L}{2\pi}\right) \tag{6.27}$$

where $\mathrm{sinc}(x) = \sin(\pi x)/(\pi x)$. For a modulator of fixed interaction length L, the group delay difference $\Delta\tau_g = \Delta\beta^{(1)}L$ reduces the accumulated phase shift with increasing modulation frequency Ω. According to Eq. 6.14, the corresponding group-delay related 3 dB-bandwidth f_{gd} is defined by $\mathrm{sinc}(f_{\mathrm{gd}}\Delta\tau_g) = 0.1410\pi/0.2500\pi$, and we find

$$f_{\mathrm{gd}} = \frac{1.7463}{\pi\Delta\tau_g}. \tag{6.28}$$

For a travelling-wave configuration where $\Delta\beta^{(1)} = 0$, the phase shift does neither degrade with electrical modulation frequency Ω, nor with device length L.

The group delay-related bandwidth f_{gd} and the RC bandwidth f_{RC} are listed in the last two columns of Table 6.1. For the strip waveguide, no RC limitation occurs. Assuming further a travelling-wave configuration and neglecting electrical dispersion, there is no inherent limitation of the electrical bandwidth. For the slot waveguide in a travelling wave configuration, the RC limitation permits operation at and above 100 Gbit/s. For the slow-light PhC modulator, the electrical signal propagates always much faster than the optical signal, and travelling-wave operation is not possible. The electrical bandwidth is therefore limited by the transit time of the optical signal in the PhC section to about 42 GHz. RC limitations can be neglected. In this case, the bandwidth does not depend on the group velocity of the optical signal – if the group velocity is further reduced, the device can be made shorter, but still the transit time remains unaffected. On the other hand, if the operating voltage U is increased, both the device length L and the transit time can be reduced. The bandwidth of the PhC modulator is therefore proportional to the operating voltage and can be extended into the 100 Gbit/s region by using larger voltage swings.

6.5 Summary

We propose novel SOI modulator schemes. Operating voltages, device lengths and modulation bandwidths are predicted. We find that 100 Gbit/s modulation at 1.5 V peak-to-peak voltage is possible with devices having a length of less than 2 mm.

Chapter 7

Third-Order Nonlinear Devices: Nonlinear SOI Waveguides for Optical Signal Processing

7.1 Introduction

Silicon-on-insulator (SOI) is considered a promising material system for dense on-chip integration of both photonic and electronic devices. Providing low absorption at infrared telecommunication wavelengths and having a high refractive index of $n \approx 3.48$, silicon is well suited for building compact linear optical devices [68, 36, 13, 111, 117, 67]. To efficiently use their inherently large optical bandwidth, it is desirable to perform all-optical signal processing and switching on the same chip by exploiting ultrafast $\chi^{(3)}$-nonlinearities such as four-wave mixing (FWM), cross and self-phase modulation (XPM, SPM) or two-photon absorption (TPA). Such devices show potential for ultrafast all-optical switching at low power [9, 34].

Third-order nonlinear interaction in SOI-based waveguides can be realized in two ways: First, nonlinear interaction with the silicon waveguide core itself can be used, leading to SPM/XPM overlaid by TPA [110, 122, 23]. Second, the silicon core can be embedded in nonlinear cladding material, and interaction with the evanescent part of the guided mode can be exploited. In the latter case, interaction with the nonlinear cladding material can be significantly enhanced when using slot waveguides rather than strips [5, 121], whereby the fraction of optical power guided in the low-index region can be maximized by appropriate waveguide dimensions [79].

However, it is not clear from the beginning, which choice leads to more pronounced nonlinearities. The strength of third-order nonlinear interaction in a waveguide is described by the nonlinear parameter γ, the real part of which depends on the waveguide geometry as well as on the nonlinear-index coefficient n_2 of the nonlinear interaction material. To optimize the waveguide dimensions for maximal nonlinear interaction, a geometrical measure is needed to rate the spatial confinement of the mode inside the nonlinear material. For optical fibers or other low index-contrast waveguides, the light propagates inside a quasi-homogeneous nonlinear material, and an appropriate measure is the so-called effective core area for nonlinear interaction A_{eff} [3] which is calculated based on a scalar approximation of the modal field. The actual cross-sectional power P related to the effective core area A_{eff} accounts then for the nonlinear deviation $n_2 P / A_{\text{eff}}$ from the linear effective refractive index of the waveguide mode.

This widely used notion of an effective area cannot be directly transferred to nonlinear high index-contrast SOI waveguides. In addition, the nonlinearity is usually limited to certain subdomains of the modal cross section.

In this chapter we therefore use the extended definition of A_{eff} introduced in Section 1.5.2. The smaller A_{eff} becomes, the larger the nonlinear effects will be for a given $\chi^{(3)}$. We calculate universal design parameters for a silicon core and for various cover materials leading to a minimum A_{eff} for strip and slot waveguides at the telecommunication wavelength $\lambda = 1.55\,\mu$m. We estimate the nonlinear waveguide parameter γ for optimized waveguide geometries. We find that γ can become more than three orders of magnitude larger ($\sim 7 \times 10^6/(\text{W km})$) than for state-of-the-art highly nonlinear fibers ($\sim 2 \times 10^3/(\text{W km})$ [66]), and we infer that ultrafast all-optical switching is feasible with non-resonant mm-scale SOI-based devices.

The chapter is structured as follows: In Section 2, we define the effective area A_{eff} for nonlinear interaction in high index-contrast waveguides; mathematical details are given in Section 1.5.2. In Section 3, we describe the waveguide optimization method, and in Section 4 we present optimal parameters for different types of SOI-based waveguides. Section 5 deals with different interaction materials; we calculate γ for various waveguides, and we discuss application examples. Section 6 summarizes the work.

Parts of the following chapter have been published in a journal article [J5], see pp. 199.

7.2 Effective Area for Third-Order Nonlinear Interaction

Figure 7.1 shows cross sections of the waveguides under consideration. The core domain D_{core} consists of silicon ($n_{\text{core}} \approx 3.48$ for $\lambda = 1.55\,\mu$m), the substrate domain D_{sub} is made out of silica ($n_{\text{sub}} \approx 1.44$), and the cover domain D_{cover} comprises a cladding material with refractive index $n_{\text{cover}} < n_{\text{core}}$. For the strip waveguide in Fig. 7.1(a), nonlinear interaction can either occur within the waveguide core ("core nonlinearity"), or the evanescent part of the guided light interacts with a nonlinear cover material ("cover nonlinearity"). The slot waveguide depicted in Fig. 7.1(b) enables particularly strong nonlinear interaction of the guided wave with the cover material inside the slot.

For maximum nonlinear interaction in strip or slot waveguides, a set of optimal geometry parameters w and h must exist: Given a nonlinear core and a linear cover, an increase of the waveguide cross section decreases the intensity inside the core and thus weakens the nonlinear interaction, while a decrease of the core size pushes the field more into the linear cover material and again reduces nonlinear effects. If a linear core is embedded into a nonlinear cover, a very small core produces a mode which penetrates the cover too deeply thus reducing the optical intensity in the nonlinear material, while for a large core only a small fraction of light will interact with the nonlinear cover.

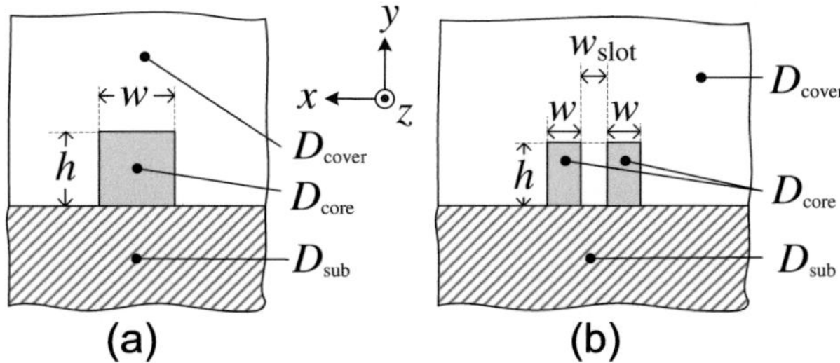

Fig. 7.1. Waveguide cross-sections (a) Strip waveguide. For core or cover nonlinearity, the nonlinear interaction domain D_{inter} is limited to the core domain ($D_{\mathrm{inter}} = D_{\mathrm{core}}$) or to the cover domain ($D_{\mathrm{inter}} = D_{\mathrm{cover}}$), respectively. (b) Slot waveguide. The nonlinear interaction domain is limited to the cover domain ($D_{\mathrm{inter}} = D_{\mathrm{cover}}$).

In Section 1.5.2 we have derived a relation for the nonlinear waveguide parameter γ which is adapted to high index-contrast waveguides, where in addition only parts of the cross section are nonlinear. We have approximated the third-order nonlinear susceptibility tensor $\widetilde{\underline{\chi}}^{(3)}$ by a scalar $\widetilde{\chi}^{(3)}$, which is constant within D_{inter}. A simple relationship of the form $\gamma \propto \widetilde{\chi}^{(3)} / \left(n_{\mathrm{inter}}^2 A_{\mathrm{eff}} \right)$ was obtained for the nonlinear waveguide parameter γ, see Eq. (1.79). In the following, the total domain $D_{\mathrm{tot}} = D_{\mathrm{core}} \cup D_{\mathrm{sub}} \cup D_{\mathrm{cover}}$ denotes the total cross section of the waveguide. D_{tot} includes a domain which is filled with the nonlinear interaction material and which is referred to as D_{inter}. The quantity n_{inter} denotes the linear refractive index of the nonlinear material in this interaction domain D_{inter}. For the case of core nonlinearity we have $D_{\mathrm{inter}} = D_{\mathrm{core}}$, $n_{\mathrm{inter}} = n_{\mathrm{core}}$, and for cover nonlinearity $D_{\mathrm{inter}} = D_{\mathrm{cover}}$, $n_{\mathrm{inter}} = n_{\mathrm{cover}}$ has to be used, see Fig. 7.1. Following Eq. (1.80), the effective area A_{eff} for third-order nonlinear interaction is then given by

$$A_{\mathrm{eff}} = \frac{Z_0^2}{n_{\mathrm{inter}}^2} \frac{\left| \iint_{D_{\mathrm{tot}}} \mathrm{Re}\left\{ \mathcal{E}_\mu(x,y) \times \mathcal{H}_\mu^*(x,y) \right\} \cdot \mathbf{e}_z \ \mathrm{d}x \ \mathrm{d}y \right|^2}{\iint_{D_{\mathrm{inter}}} \left| \mathcal{E}_\mu(x,y) \right|^4 \mathrm{d}x \ \mathrm{d}y}. \tag{7.1}$$

The subscript μ denotes the waveguide mode, and we have skipped the frequency arguments ω_{c} of the mode fields. $Z_0 = \sqrt{\mu_0/\epsilon_0} = 377\,\Omega$ is the free-space wave impedance, and $\mathbf{e}_z$ is the unit vector pointing in positive z-direction. For low-index contrast material systems with homogeneous nonlinearity, Eqs. (1.80) and (7.1) reduce to the usual definition of an effective area [3, Eq. (2.3.29)] as is shown in Eq. (1.81).

The modal fields $\mathcal{E}_\mu(x,y)$ and $\mathcal{H}_\mu(x,y)$ are classified by the terms TE and TM. TE refers to a waveguide mode with a dominant electric field component in x-direction (parallel to the substrate plane), whereas the dominant electric field component of a TM mode is directed parallel to the y-axis (perpendicular to the substrate plane).

7.3 Waveguide Optimization Method

To evaluate the integrals in Eq. (7.1), both the electric and the magnetic fields of the fundamental waveguide modes are calculated using a commercially available vectorial finite-element mode solver [92]. For core (cover) nonlinearity, the computational domain extends from $-1.5\,\mu$m to $+1.5\,\mu$m ($-2\,\mu$m to $+2\,\mu$m) in the x-direction, and from $-1\,\mu$m to $+2\,\mu$m ($-1.5\,\mu$m to $+2.5\,\mu$m) in the y-direction, terminated by perfectly matched layers of $0.4\,\mu$m thickness in all directions. To improve accuracy, second-order finite elements are used. The size of the finite elements outside the core region is $\Delta x \approx \Delta y \approx 40\,$nm, whereas the silicon strips and the gaps are each divided into at least 10 elements both in the x- and in the y-direction. To better resolve the discontinuities of the normal electric field components, two layers of $2\,$nm wide finite elements are placed on each side of each dielectric interface. For the structures operated in TM polarization, the fields are evaluated and stored on a rectangular grid with step size $\Delta x_\mathrm{store} \approx 5\,$nm in the x-direction and $\Delta y_\mathrm{store} \approx 2\,$nm in the y-direction. For TE polarization, the values $\Delta x_\mathrm{store} \approx 2\,$nm in x-direction and $\Delta y_\mathrm{store} \approx 5\,$nm in y-direction are chosen. The exact step sizes of the grids are matched to hit the dielectric boundaries.

For optimization, the waveguide parameters w and h are alternately scanned in a certain range. The resulting values for A_eff are slightly scattered due to numerical inaccuracies. Therefore, a fourth-order polynomial is fitted to the data points, and the local minimum of the polynomial is taken as a starting point for the next scan. The iteration is stopped when the geometrical parameters repeatedly change by less than $0.5\,$nm between subsequent iterations.

7.4 Optimal Strip and Slot Waveguides

Third-order nonlinear interaction is maximized for five different cases: Core nonlinearities in strip waveguides for both TE- and TM-polarization, cover nonlinearities in strip waveguides for both polarizations, and cover nonlinearities in TE-operated slot waveguides. For the exploitation of core (cover) nonlinearities, different values of $n_\mathrm{cover} \in \{1.0, 1.1, \ldots 2.5\}$ ($n_\mathrm{cover} \in \{1.0, 1.1, \ldots 3.0\}$) are considered.

7.4.1 Strip Waveguides and Core Nonlinearity

For the case of *core nonlinearity*, silicon is used as nonlinear interaction material. Silicon is of point group $m3m$. If Kleinman's symmetry is assumed, the susceptibility tensor has two independent elements, $\widetilde{\chi}^{(3)}_{1111} = \widetilde{\chi}^{(3)}_{2222} = \widetilde{\chi}^{(3)}_{3333}$ and $\widetilde{\chi}^{(3)}_{1122} = \widetilde{\chi}^{(3)}_{1212} = \widetilde{\chi}^{(3)}_{1221} = \widetilde{\chi}^{(3)}_{2211} = \cdots = \widetilde{\chi}^{(3)}_{1133} = \cdots = \widetilde{\chi}^{(3)}_{2233} = \ldots$, where the indices 1, 2 and 3 refer to the crystallographic [100], [010] and [001] directions. For an isotropic nonlinearity, $\widetilde{\chi}^{(3)}_{1122}/\widetilde{\chi}^{(3)}_{1111} = 1/3$, but for silicon a larger ratio $\widetilde{\chi}^{(3)}_{1122}/\widetilde{\chi}^{(3)}_{1111} = 0.48 \pm 0.03$ has been measured [120]. The assumption of an anisotropic nonlinearity is thus not valid in the strict sense and implies that the components of the nonlinear polarization vector that are not oriented parallel to the exciting electric field vector are neglected. However, the error in calculating the nonlinear waveguide parameter γ is negligible: The TM (TE) mode fields have a domi-

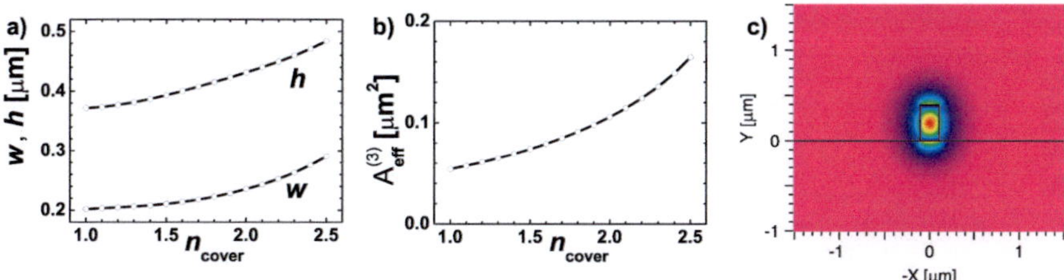

Fig. 7.2. TM-operated strip waveguide with *core nonlinearity*. Optimized geometrical parameters for a minimum effective area A_{eff} (a) Optimal strip width w and height h as a function of the refractive index n_{cover} of the linear cover material (b) Minimized effective area A_{eff} of nonlinear interaction. (c) Dominant component $(\mathcal{E}_{\mu\,y})$ of the electric modal field for $n_{\text{cover}} = 1.5$

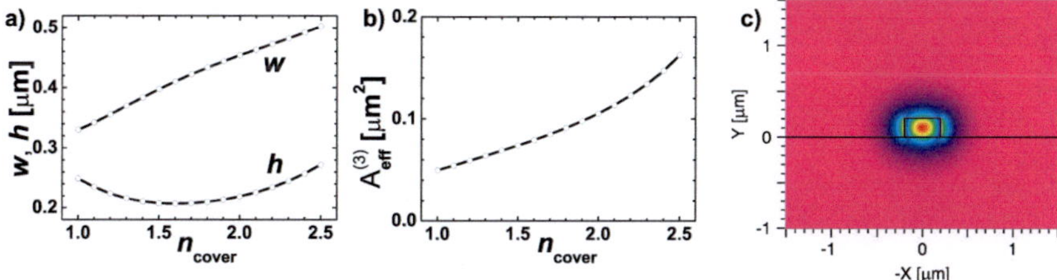

Fig. 7.3. TE-operated strip waveguide with *core nonlinearity*. Optimized geometrical parameters for a minimum effective area A_{eff} (a) Optimal strip width w and height h as a function of the refractive index n_{cover} of the linear cover material (b) Optimized effective area A_{eff} of nonlinear interaction (c) Dominant component $(\mathcal{E}_{\mu\,x})$ of the electric modal field for $n_{\text{cover}} = 1.5$

nant $\mathcal{E}_{\mu\,y}$-component ($\mathcal{E}_{\mu\,x}$-component), resulting, e.g., in an inaccurate x-component (y-component) of the nonlinear polarization. To calculate the overlap integral in Eq. (1.78) these components are weighted with the weak $\mathcal{E}_{\mu\,x}$-component ($\mathcal{E}_{\mu\,y}$-component) for TM (TE). The overall error is thus very small compared to the contributions of the non-linear polarization's y-component (x-component). The error in γ would increase, if the interaction between modes of orthogonal polarizations was of interest: The nonlinear polarization generated by a TM (TE) mode is then projected onto a TE (TM) mode field. A small, but inaccurate x-component (y-component) of the nonlinear polarization is thus weighted with the dominant component $\mathcal{E}_{\mu\,x}$-component ($\mathcal{E}_{\mu\,y}$-component), whereas the large y-component (x-component) of the nonlinear polarization is weighted by the weak $\mathcal{E}_{\mu\,x}$-component ($\mathcal{E}_{\mu\,y}$-component). However, from a practical point of view, theses inaccuracies are small compared to the uncertainties in measured nonlinearities of silicon, Table 7.1.

Figure 7.2 shows the results for *core nonlinearity* in a TM-operated strip waveguide. The dominant electric field component $(\mathcal{E}_{\mu\,y})$ is discontinuous at the horizontal dielectric interfaces with a strong field enhancement in the low-index material. Therefore the optimal cross sectional shape of the waveguide core must be narrow and high. This is

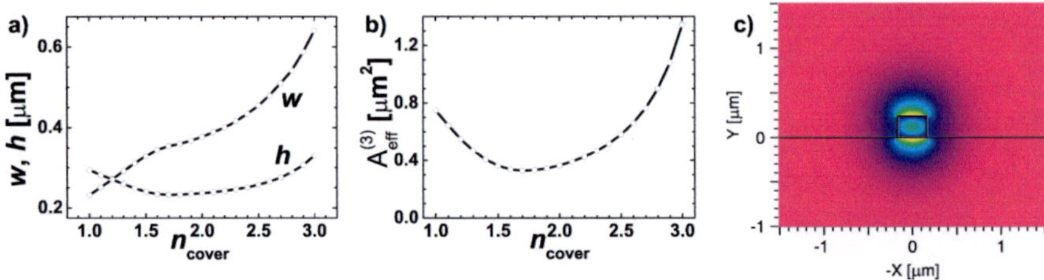

Fig. 7.4. TM-operated strip waveguide with *cover nonlinearity*. Optimized geometrical parameters for a minimum effective area A_{eff} (a) Optimal strip width w and height h as a function of the linear refractive index n_{cover} of the nonlinear cover material (b) Minimized effective area A_{eff} of nonlinear interaction (c) Dominant component $(\mathcal{E}_{\mu\,y})$ of the electric modal field for $n_{\text{cover}} = 1.5$

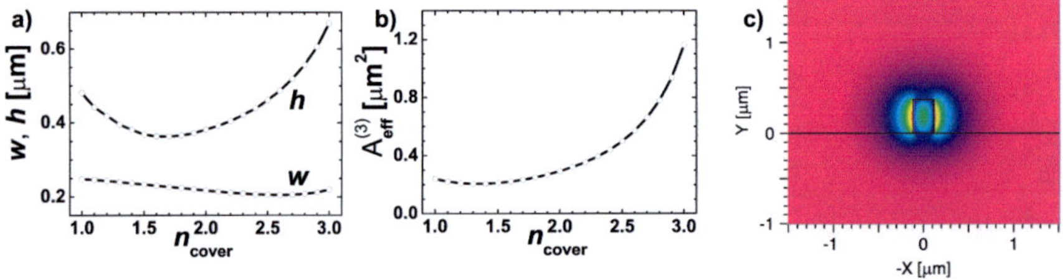

Fig. 7.5. TE-operated strip waveguide with *cover nonlinearity*. Optimized geometrical parameters for a minimum effective area A_{eff} (a) Optimal strip width w and height h as a function of the linear refractive index n_{cover} of the nonlinear cover material (b) Minimized effective area A_{eff} of nonlinear interaction (c) Dominant component $(\mathcal{E}_{\mu\,x})$ of the electric modal field for $n_{\text{cover}} = 1.5$

confirmed by the results of the optimization. It can further be seen that a high index contrast between the core and the cover material always allows for higher field confinement and stronger nonlinear interaction within the core. Effective nonlinear interaction areas as small as $A_{\text{eff}} = 0.054\,\mu\text{m}^2$ can be obtained for $n_{\text{cover}} = 1.0$.

Figure 7.3 shows the results for *core nonlinearity* in a TE-operated strip waveguide. Using analogous arguments as for the TM case, the optimal cross section of the waveguide core must now be wide and flat. Again, a high index contrast between the core and the cover material always allows for higher field confinement and stronger nonlinear interaction within the core. For low values of n_{cover}, the minimal effective area of nonlinear interaction is slightly smaller for TE polarization than it was TM — for $n_{\text{cover}} = 1.0$ we now find $A_{\text{eff}} = 0.050\,\mu\text{m}^2$. TE-operated strip waveguides with silica cover ($n_{\text{cover}} = 1.44$) and with nearly optimal width $w = 400\,\text{nm}$ and height $h = 200\,\text{nm}$ have previously been used in experiments [111, 37].

7.4.2   Strip Waveguides and Cover Nonlinearity

The results for *cover nonlinearity* in TM-operated strip waveguides are shown in Fig. 7.4. The dominant electric field component $(\mathcal{E}_{\mu\,y})$ is discontinuous at horizontal dielectric interfaces with a strong field enhancement in the nonlinear low-index material. Under these circumstances, the optimal cross sectional shape of the waveguide is rather wide and flat except for very low refractive indices of the cladding material. It is further found that there is an optimal refractive index $n_{\text{cover}} \approx 1.7$ for which A_{eff} assumes a minimal value of $0.33\,\mu\text{m}^2$. For lower indices, too big a fraction of the electromagnetic field has to be guided within the waveguide core to prevent leakage into the substrate. This part of the field does not contribute to the nonlinear interaction, which makes the effective area bigger. For higher refractive indices, the field enhancement at the dielectric interface decreases, which reduces the nonlinear interaction with the cover material.

In the case of a TE-operated strip waveguide with *cover nonlinearity*, discontinuous field enhancement can be exploited at both sidewalls. This results in smaller effective nonlinear interaction areas as can be seen from Fig. 7.5. The minimum of A_{eff} now shifts to $n_{\text{cover}} \approx 1.3$ and amounts to roughly $0.24\,\mu\text{m}^2$.

7.4.3 Slot Waveguides and Cover Nonlinearity

For a slot waveguide, most of the light is confined to the slot area, and reducing the slot width w_{slot} increases the intensity in the nonlinear material. Within the range of technologically feasible slot widths, the effective nonlinear interaction area A_{eff} therefore always decreases with w_{slot} and no optimal value for w_{slot} can be found. For the design of slot-waveguides, the minimum slot width will be dictated by technological issues, e.g. the maximum aspect ratio that the fabrication process can achieve, or the difficulty of filling a narrow slot with nonlinear interaction material. Therefore $w_{\text{slot}} \in \{60\,\text{nm},\,80\,\text{nm}\,\ldots\,200\,\text{nm}\}$ is fixed during the optimization procedure.

Figure 7.6 shows the optimal parameters as a function of the refractive index n_{cover} of the nonlinear cover material with the slot width w_{slot} as a parameter. The width w of the individual strips mainly depends on n_{cover}, whereas the optimal height h shows substantial variations with both n_{cover} and w_{slot}. For $w_{\text{slot}} \geq 100\,\text{nm}$, there is again an optimal refractive index n_{cover} for which A_{eff} is minimum. The existence of this minimum can be explained physically: For larger refractive indices, the discontinuity-induced field enhancement at the dielectric interfaces decreases. For lower refractive indices, the increase in field enhancement is over-compensated by the fact that a minimum fraction of the electromagnetic field has to be guided in the high-index core material to prevent leakage into the substrate. This fraction of the field does not contribute to the nonlinear interaction and thus increases A_{eff}. For $w_{\text{slot}} < 100\,\text{nm}$, the guidance of the fundamental mode is always strong enough to prevent it from leaking into the substrate, and A_{eff} decreases monotonically as n_{cover} decreases.

Similar arguments hold for explaining the behaviour of the optimal height: For decreasing refractive indices, the height increases in the case of $w_{\text{slot}} \geq 120\,\text{nm}$ to prevent leakage into the substrate. For $w_{\text{slot}} < 120\,\text{nm}$ this does not seem to be crucial, and the optimal height even decreases slightly for small values of n_{cover}. Using slot waveguides with technologically feasible gap widths of $100\,\text{nm}$ results in effective nonlinear interaction areas as small as $A_{\text{eff}} = 0.086\,\mu\text{m}^2$ or $A_{\text{eff}} = 0.105\,\mu\text{m}^2$ for $n_{\text{cover}} = 1.2$ or $n_{\text{cover}} = 1.5$, respectively.

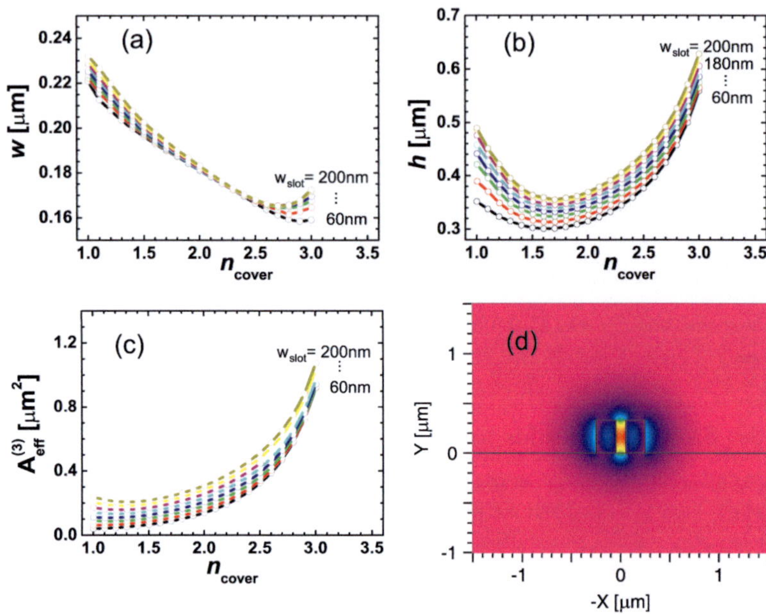

Fig. 7.6. TE-operated slot waveguide with *cover nonlinearity*. Optimized geometrical parameters for a minimum effective area A_{eff} (a) Optimal strip width w as a function of the linear refractive index n_{cover} of the nonlinear cover material for various slot widths $w_{\text{slot}} \in \{60\,\text{nm}, 80\,\text{nm}, \dots, 200\,\text{nm}\}$ (b) Optimal strip height h (c) Minimized effective area A_{eff} for nonlinear interaction   (d) Dominant component $(\mathcal{E}_{\mu\,x})$ of the electric modal field for $n_{\text{cover}} = 1.5$ and $w_{\text{slot}} = 100\,\text{nm}$.

7.5   Nonlinear Parameters for Different Materials

The previous analysis shows that outstandingly small effective areas A_{eff} can be obtained in SOI-based waveguides, and it can be expected that, depending on the properties of the employed materials, highly nonlinear integrated waveguides can be realized. We will now estimate the nonlinear parameter $\operatorname{Re}\{\gamma\}$ for different interaction materials.

Nonlinear properties of optical materials are commonly described by a nonlinear refractive index which depends on the intensity I of an optical wave, $n = n_0 + n_2 I$, and by a corresponding intensity-dependent power absorption coefficient $\alpha = \alpha_0 + \alpha_2 I$. The nonlinear refractive index n_2 and the TPA coefficient α_2 are linked to the scalar third-order nonlinear optical susceptibility $\widetilde{\chi}^{(3)}$ by [3, Eq. (2.3.13)]

$$n_2 = \frac{3Z_0 \operatorname{Re}\{\widetilde{\chi}^{(3)}\}}{4n_0^2}, \tag{7.2}$$

$$\alpha_2 = -\frac{3k_0 Z_0 \operatorname{Im}\{\widetilde{\chi}^{(3)}\}}{2n_0^2}. \tag{7.3}$$

| Material | Re $\{\gamma\}/(\mathrm{W\,m})^{-1}$ | | $\lambda/\,\mathrm{nm}$ | n_0 | $n_2/(\mathrm{m}^2/\mathrm{W})$ | $\mathrm{FOM_{TPA}}$ | Ref. |
	$\mathrm{TM_{strip}}$	$\mathrm{TE_{strip}}$					
Silicon	449	487	1 550	3.48	6×10^{-18}	0.86	[110]
	336	365	1 540	3.48	4.5×10^{-18}	0.37	[22]
	322	349	1 540	3.48	4.3×10^{-18}	0.32	[22]
	1 080	1 180	1 550	3.48	14.5×10^{-18}	1.56	[122]
	374	406	1 550	3.48	5×10^{-18}		[23]

Table 7.1. Core nonlinearity. Calculated maximum nonlinearity parameters $\mathrm{Re}\{\gamma\} \propto 1/A_{\mathrm{eff}}$ for optimized strip waveguides with a nonlinear silicon core and a linear air cladding $n_{\mathrm{cover}} = 1$, operated in TM or TE polarization. The calculation is based on data for silicon at the specified wavelengths: Linear refractive index n_0, nonlinearity coefficient n_2 and TPA figure of merit $\mathrm{FOM_{TPA}}$ were taken from the references listed in the last column. — The resulting nonlinear parameters $\mathrm{Re}\{\gamma\} \approx 400/(\mathrm{W\,m})$ are remarkably large. However, the material suffers from non-negligible two-photon absorption leading to a figure of merit $\mathrm{FOM_{TPA}} \approx 0.3\ldots 0.9$.

TPA leads to a strong decay of optical power along the direction of propagation and can therefore severely impair nonlinear parametric effects such as SPM, XPM and FWM [77]. A measure of this impairment is the TPA figure of merit $\mathrm{FOM_{TPA}}$, which is the nonlinear phase shift related to the associated intensity change and may be expressed through the nonlinear parameter γ, see Eq. (1.79),

$$\mathrm{FOM_{TPA}} = -\frac{1}{2\pi}\frac{\mathrm{Re}\{\gamma\}}{2\,\mathrm{Im}\{\gamma\}} = \frac{n_2}{\alpha_2\lambda}. \tag{7.4}$$

An optical power P_0 launched into a waveguide of length L would account for a nonlinear phase shift of $\Delta\phi_{\mathrm{nl}} = \mathrm{Re}\{\gamma\}\,P_0 L$ in the absence of loss. TPA reduces the power along the propagation length, $P(L) = P_0/(1 + \frac{\Delta\phi_{\mathrm{nl}}}{2\pi}\mathrm{FOM_{TPA}})$, thereby reducing the nonlinear phase shift. To achieve SPM-induced nonlinear phase-shifts $\Delta\phi_{\mathrm{nl}} > 2\pi$ $(\Delta\phi_{\mathrm{nl}} > \pi)$, the interaction material should satisfy $\mathrm{FOM_{TPA}} > 1$ $(\mathrm{FOM_{TPA}} > 0.5)$ [44].

Tables 7.1 and 7.2 list the calculated optimum nonlinear parameters $\mathrm{Re}\{\gamma\}$ as defined in Eq. (1.79) for various nonlinear core and cover materials, polarizations and structures. In both tables these calculations are based on material data at the specified wavelengths, namely on the linear refractive index n_0 and on the nonlinearity coefficient n_2. In addition, the TPA figure of merit $\mathrm{FOM_{TPA}}$ is specified. All material data were taken from the references listed in the last column of both tables. For some materials, no $\mathrm{FOM_{TPA}}$ data at 1 550 nm could be found. Some nonlinearity data were only available from third-harmonic generation experiments, which is indicated in Table 7.2 by an asterisk$^{\star)}$ after the wavelength. In these cases the calculated maximum nonlinear parameter $\mathrm{Re}\{\gamma\}$ might be inaccurate, but should still reflect the correct order of magnitude.

Table 7.1 refers to the case of *core nonlinearity* with silicon as the nonlinear core material. Reported nonlinearity coefficients n_2 for silicon range from $4.3 \times 10^{-18}\,\mathrm{m}^2/\mathrm{W}$ to $14.5 \times 10^{-18}\,\mathrm{m}^2/\mathrm{W}$. The nonlinear parameters $\mathrm{Re}\{\gamma\}$ have been calculated for optimized strip waveguides with air as a cover material $(n_{\mathrm{cover}} = 1.0)$. Optimal strip widths and heights for TM-polarization, ("$\mathrm{TM_{strip}}$", $A_{\mathrm{eff}} = 0.054\,\mu\mathrm{m}^2$) and for TE-polarization ("$\mathrm{TE_{strip}}$", $A_{\mathrm{eff}} = 0.050\,\mu\mathrm{m}^2$) are obtained from Figs. 7.3 and 7.2. Depending on the value of n_2, the resulting nonlinear waveguide parameters range from $322/(\mathrm{W\,m})$ to

Material	$\mathrm{Re}\{\gamma\}/(\mathrm{W\,m})^{-1}$			$\lambda/\,\mathrm{nm}$	n_0	$n_2/(\mathrm{m}^2/\mathrm{W})$	$\mathrm{FOM_{TPA}}$	Ref.
	$\mathrm{TM_{strip}}$	$\mathrm{TE_{strip}}$	$\mathrm{TE_{slot}}$					
Inorganic materials								
Pure silica glass	0.3	0.5	1.0	1 550	1.45	2.48×10^{-20}	$\gg 10$	[3]
Lead silicate glass								
Schott SF59	8.0	11	17	1 060	1.91	6.8×10^{-19}		[28]
Bismite glass	3.5	4.4	6.9	1 550	2.02	3.2×10^{-19}		[56]
Tellurite glass								
$\mathrm{Li_2O\text{-}TiO_2\text{-}TeO_2}$	6.2	7.5	11	$1\,900^{\star)}$	2.2	6.53×10^{-19}		[80]
Chalcogenide glass								
$\mathrm{As_{24}S_{38}Se_{38}}$	74	86	120	1 600	2.45	1.0×10^{-17}		[16]
$\mathrm{As_{39}Se_{61}}$	71	82	105	1 500	2.81	1.6×10^{-17}	3.8	[47]
$\mathrm{As_{40}Se_{60}}$	102	117	151	1 500	2.81	2.3×10^{-17}	11	[47]
Organic materials								
PDA	54	92	186	1 319	$\sim$1.5	4.8×10^{-18}	1.5	[109, 89]
PTA	22	38	78	$1\,907^{\star)}$	1.5	2×10^{-18}		[44]
TEE	17	29	58	$1\,907^{\star)}$	1.5	1.5×10^{-18}		[43]
PSTF66	31	54	109	1 550	1.5	2.8×10^{-18}	0.22	[52, 7]
DANS	94	149	293	1 319	1.57	8×10^{-18}	7.6	[57]
PTS (PDA)	2 720	3 820	6 950	1 600	$\sim$1.7	2.2×10^{-16}	>27	[12, 62]
Nanocomposites								
Si nanocrystals	22 800	33 100	61 600	813	1.66	1.86×10^{-15}	5.6	[115]
	1 120	1 910	4 000	1 500	1.5	$\sim10^{-16}$		[9]

$^{\star)}$ Third-order nonlinearity obtained by third-harmonic generation

Table 7.2. Cover nonlinearity. Calculated maximum nonlinearity parameters $\mathrm{Re}\{\gamma\}\propto 1/A_{\mathrm{eff}}$ for optimized strip and slot waveguides with a linear silicon core and various nonlinear cover materials, operated in TM or TE polarization. The calculation is based on cover material data at the specified wavelengths: Linear refractive index n_0, nonlinearity coefficient n_2 and TPA figure of merit $\mathrm{FOM_{TPA}}$ were taken from the references listed in the last column. Three material groups are considered: Inorganic materials like glasses, organic substances, and nanocomposites. — Most remarkable are the large nonlinear parameters $\mathrm{Re}\{\gamma\}\approx(70\ldots150)/(\mathrm{W\,m})$ and $\mathrm{Re}\{\gamma\}\approx 300/(\mathrm{W\,m})$ for chalcogenide glasses and for the side-chain polymer DANS, respectively, and the record value of $\mathrm{Re}\{\gamma\}\approx 7\,000/(\mathrm{W\,m})$ for the single-crystalline organic material PTS, a number which is 1 000 times larger than for a higly nonlinear bismite glass. These material groups have also very good TPA figures of merit in the order of $\mathrm{FOM_{TPA}}\approx 4\ldots 27$.

$1\,180/(\mathrm{W\,m})$. TPA figures of merit around 1 indicate that parametric effects such as SPM, XPM and FWM will usually be impaired by TPA.

Table 7.2 refers to the case of *cover nonlinearity*. The interaction material must have a linear refractive index $n_{\mathrm{inter}}=n_0$ smaller than the index of silicon and provide low linear and nonlinear absorption in the desired wavelength range. There is a vast choice of such materials, and we have concentrated on the most prominent ones for which reliable data on nonlinear parameters could be obtained. These materials are subdivided into three groups: Inorganic materials (glasses), organic materials (polymers) and nanocomposites (e.g., artificial nanocrystals).

For each material, we have estimated the nonlinear parameter $\mathrm{Re}\{\gamma\}$ for three different cases: A TM-operated strip waveguide ("$\mathrm{TM_{strip}}$"), a TE-operated strip waveguide ("$\mathrm{TE_{strip}}$"), and a TE-operated slot waveguide with $w_{slot} = 100\,\mathrm{nm}$ ("$\mathrm{TE_{slot}}$"). All these waveguides have geometries optimized for the respective cover material, see Figs. 7.4, 7.5 and 7.6. The nonlinear parameter $\mathrm{Re}\{\gamma\}$ denotes the contribution of the nonlinear cover material only — the contribution of the silicon core is not taken into account, and the values for $\mathrm{Re}\{\gamma\}$ as listed in Tab. 7.2 are to be understood as lower bounds for the nonlinear parameter. While the waveguides discussed in Tab. 7.1 are designed with a nonlinear core material, the structures in Tab. 7.2 have been optimized for cover nonlinearity; the contribution of the silicon core is in this case significantly smaller than could be inferred from Tab. 7.1.

The first group of nonlinear cover materials comprises different glasses. Silica glass ($\mathrm{SiO_2}$) is not a typical nonlinear material, but for comparison, we have calculated the corresponding nonlinear parameters. We note that the resulting values $\mathrm{Re}\{\gamma\} \leqslant 1.0/(\mathrm{W\,m})$ are in the same order of magnitude as the nonlinear parameters obtained for modern highly-nonlinear fibers based on lead silicate glasses, $\mathrm{Re}\{\gamma\} = 1.86/(\mathrm{W\,m})$ [66]. Lead silicate glasses, bismite glasses, tellurite glasses and chalcogenide glasses feature high linear and high nonlinear refractive indices n_0 and n_2. The high linear indices considerably reduce the discontinuity-induced field enhancement at the dielectric interfaces, so that A_{eff} increases and the nonlinear parameter decreases. For the slot waveguide, we find $A_{\mathrm{eff}} = 0.62\,\mu\mathrm{m}^2$ given $n_{\mathrm{cover}} = 2.81$, which is roughly a factor of 6 bigger than the value of $A_{\mathrm{eff}} = 0.104\,\mu\mathrm{m}^2$ for $n_{\mathrm{cover}} = 1.5$. Still, the nonlinear parameters $\mathrm{Re}\{\gamma\}$ are nearly two orders of magnitude larger than for state-of-the-art highly nonlinear fibers [66].

The second group of nonlinear materials comprises nonlinear organic materials. Nonlinearities in these materials can either arise from the polymer backbone, or from chromophore units embedded in the host matrix or laterally attached to the backbone. For the conjugated polymers PDA (polydiacetylene), PTA (polytriactelyene) and TEE (tetraethynylethene), nonlinearities are roughly two orders of magnitude stronger than for $\mathrm{SiO_2}$. Please note that the nonlinear refractive indices for PTA and TEE have been measured via third-harmonic generation (THG) at a pump wavelength of $1\,900\,\mathrm{nm}$, and the results cannot offhand be applied to SPM at $1\,550\,\mathrm{nm}$. The order of magnitude might be correct, though. The organic dye functionalized main-chain polymer PSTF66 exhibits large nonlinear losses, whereas the side chain polymer DANS (4-dialkyamino-4'nitro-stilbene) exhibits TPA figures of merit that are suitable for devices based on nonlinear phase shifts. For single-crystalline poly(p-toluene sulphonate) (PTS) polydiacetylene, nonlinear refraction is even four orders of magnitude stronger than for $\mathrm{SiO_2}$, and nonlinear parameters $\mathrm{Re}\{\gamma\}$ in the order of $6\,950/(\mathrm{W\,m})$ can be expected for slot waveguides without severe impairment by TPA. For strip waveguides, $\mathrm{Re}\{\gamma\}$ reduces by roughly 50 %, but is still about $3\,820/(\mathrm{W\,m})$. Using single-crystal PTS as a nonlinear interaction material around a pre-structured silicon waveguide core might also solve the problem of poor processability of single crystal PTS.

Lastly, we consider the case where the slot waveguide is filled with artificial silicon nanocrystals. At $\lambda = 813\,\mathrm{nm}$ this nanocomposite material exhibits huge nonlinearities (about five order of magnitudes stronger than in $\mathrm{SiO_2}$) without impairment by TPA. It is questionable which nonlinearities can be obtained at $1\,550\,\mathrm{nm}$, but even if only values of $n_0 = 1.50$ and $n_2 = 10^{-16}\,\mathrm{m}^2\,/\,\mathrm{W}$ are assumed, as has been done by other authors [9], large nonlinear parameters $\mathrm{Re}\,\{\gamma\}$ up to $4\,000/\,(\mathrm{W\,m})$ can be expected.

7.6 Application Examples

For state-of-the-art highly nonlinear fibers, the highest nonlinear parameters $\mathrm{Re}\,\{\gamma\}$ are in the order of $2/\,(\mathrm{W\,m})$ [66]. According to our estimations, a nonlinear parameter more than three orders of magnitude larger can be expected for SOI-based strip and slot waveguides covered with appropriate nonlinear interaction materials. Approximately one order of magnitude is gained from the strong confinement of the electromagnetic field. Because waveguides with cover nonlinearities allow to choose from a broad spectrum of interaction materials, the extremely nonlinear PTS can be chosen, which leads to an additional improvement of approximately two orders of magnitude compared to lead silicate glass.

Highly-nonlinear integrated strip and slot waveguides are viable for on-chip all-optical signal processing as shall be illustrated by estimating the lengths required for a passive SPM/XPM-based switch and a passive wavelength converter based on FWM.

The nonlinear phase shift $\Delta\phi_{\mathrm{nl}}$ experienced by an optical signal through SPM or XPM in a lossless waveguide is proportional to the optical power P and the interaction length L, $\Delta\phi_{\mathrm{nl}} = \mathrm{Re}\,\{\gamma\}\,PL$ or $\Delta\phi_{\mathrm{nl}} = 2\,\mathrm{Re}\,\{\gamma\}\,PL$, respectively. For many nonlinear signal processing schemes, a nonlinear phase shift of $\Delta\phi_{\mathrm{nl}} = \pi$ is required. If an optical peak power of $P = 100\,\mathrm{mW}$ and a slot waveguide with a nonlinear waveguide parameter of $\mathrm{Re}\,\{\gamma\} = 6\,950/\,(\mathrm{W\,m})$ are assumed, a nonlinear phase shift of π requires a slot waveguide with a length of $L = 4.5\,\mathrm{mm}$ or $L = 2.3\,\mathrm{mm}$, respectively. For $\mathrm{Re}\,\{\gamma\} = 3\,820/\,(\mathrm{W\,m})$ as calculated for a TE-operated strip waveguide, the length increases to $L = 8.2\,\mathrm{mm}$ or $L = 4.1\,\mathrm{mm}$, again for SPM or XPM, respectively.

Neglecting waveguide loss and pump depletion, and assuming phase matching, the conversion efficiency for degenerate FWM is given by $\eta_{\mathrm{FWM}} = (\mathrm{Re}\,\{\gamma\}\,P_{\mathrm{pmp}}L)^2$, where P_{pmp} denotes the pump power [3]. Assuming again a slot waveguide with $\mathrm{Re}\,\{\gamma\} = 6\,950/\,(\mathrm{W\,m})$ and $P_{\mathrm{pmp}} = 100\,\mathrm{mW}$, a conversion efficiency of $100\,\%$ can be obtained for an estimated waveguide length of $L = 1.4\,\mathrm{mm}$. For a TE-operated strip waveguide with $\mathrm{Re}\,\{\gamma\} = 3\,820/\,(\mathrm{W\,m})$, this length increases to $L = 2.6\,\mathrm{mm}$.

These results indicate that broadband, i.e., nonresonant ultrafast all-optical signal processing is feasible with compact mm-long integrated devices based on highly nonlinear slot and strip waveguides. We note that in all cases the assumed power levels are far too low to induce saturation of the nonlinear phase shift due to a Kerr-induced decrease of the discontinuity-induced field enhancement [35]. As with all nonlinear switching processes, the switching power and/or the interaction length can be considerably reduced at the expense of bandwidth by using resonant structures [9]. Compared to signal processing schemes based on active integrated devices, e.g., semiconductor optical amplifiers, passive schemes need higher power levels. However, passive Kerr-based devices are ultra-fast, do not exhibit pattern effects, and do not require active cooling.

7.7 Summary

SOI-based nonlinear strip and slot waveguides are well suited for ultrafast all-optical signal processing if an appropriate cover material is applied. A newly introduced effective area A_{eff} for third-order nonlinear interaction in high index-contrast waveguides with nonlinear constituents serves as a basis for the optimization of different SOI-based waveguide structures with respect to a maximum nonlinearity parameter γ. We provide universal optimal design parameters for strip and slot waveguides covered with different nonlinear interaction materials, and we calculate the resulting maximum nonlinear parameter γ. It is found that γ can be more than three orders of magnitude larger compared with state-of-the-art highly nonlinear fibers. Estimating the waveguide lengths for different nonlinear signal processing schemes, we infer that nonresonant ultrafast nonlinear signal processing is possible with mm-scale integrated SOI-based devices.

Chapter 8

Nonlinear Dynamics: Gain and Phase Response of an InAs/GaAs Quantum Dot amplifier at 1300 nm

8.1   Introduction

Self assembled quantum dot (QD) semiconductor optical amplifiers (SOA) are promising devices for pattern-effect-free linear and nonlinear all-optical signal processing [104]. Large unsaturated gain and large output saturation power can be obtained over a broad spectral range of more than 100 nm [4] while maintaining a low noise figure that can approach the limit of 3 dB [11]. Featuring sub-10 ps carrier recovery times, QD semiconductor optical amplifier (SOA) enable all-optical signal processing at high data rates, whereby the wetting layer acts as a reservoir for carriers. The interaction between multiple channels can be significantly reduced if their wavelength separation is larger than the homogeneous bandwidth of the single-dot gain. This allows for multi-wavelength signal processing within a single device. On the other hand, strong nonlinear interaction via the cross-gain modulation (XGM) mechanism can take place when the input channels are located within the homogenous line broadening [103]. In addition, InAs/GaAs QD SOA exhibit low sensitivity with respect to temperature and show potential for uncooled operation at 1 300 nm.

To design SOA-based nonlinear all-optical switches or wavelength converters, the magnitudes and dynamics of gain and phase changes must be known. Pump-probe measurements of these parameters have been done for QD SOA both in the 1 300 nm and in the 1 550 nm wavelength region, see [124] and the references therein. Yet a detailed investigation of the relative magnitudes of gain and phase changes in InAs/GaAs QD SOA has still to be done.

In this chapter, we measure sub-10 ps dynamics and magnitudes of gain and phase changes in such devices, and we investigate their suitability for different signal processing schemes. We find that pump-induced gain saturation dominates over nonlinear phase shifts, and we infer that QD SOA are especially useful in signal processing schemes that exploit cross-gain modulation.

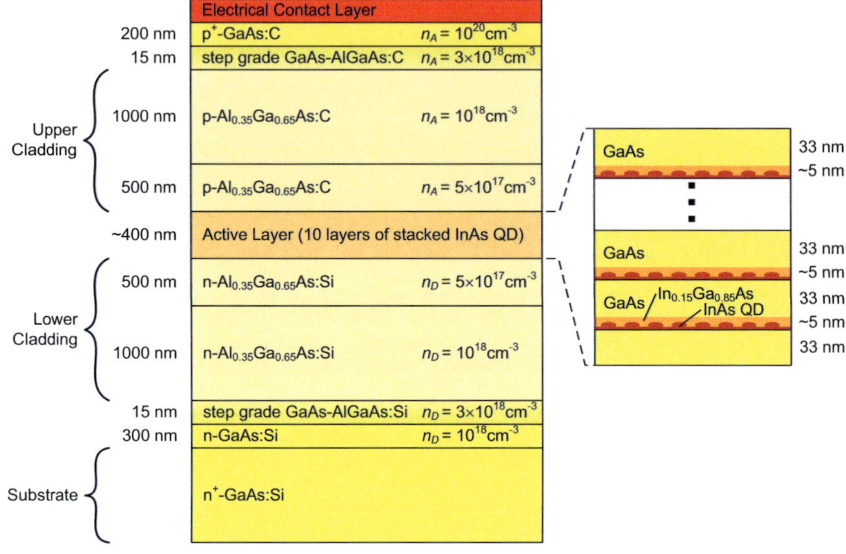

Fig. 8.1. Epitaxial structure of the quantum dot semiconductor optical amplifier. The quantity n_D (n_A) denotes the donor (acceptor) density in the respective n-doped (p-doped) layers.

8.2 Semiconductor Optical Amplifier Structure

8.2.1 Epitaxial Layers

The quantum dots were grown with molecular beam epitaxy (MBE) on an n-doped GaAs (100) substrate. Fig. 8.1 shows the epitaxial structure. The active region consists of 10 stacked layers of self-assembled InAs quantum dots, each covered with a 5 nm thick $\mathrm{In_{0.15}Ga_{0.85}As}$ layer and a GaAs spacer of 33 nm thickness. In each layer, the QD density is approximately $1.3 \times 10^{11}\,\mathrm{cm^{-2}}$. The active region is surrounded by $1.5\,\mu$m thick p- and n-doped $\mathrm{Al_{0.35}Ga_{0.65}As}$ cladding layers above and below, respectively. For detailed information on the growth process of the quantum dots, refer to [59].

8.2.2 Active Waveguide

The investigated SOA (device DO520_04a) consists of a $L = 2$ mm long deeply etched ridge waveguide of $4\,\mu$m width. The etch depth is typically $2\,\mu$m. The waveguide cross section is depicted in Fig. 8.2. For a detailed description of the processing steps, refer to [61, Device DO520].

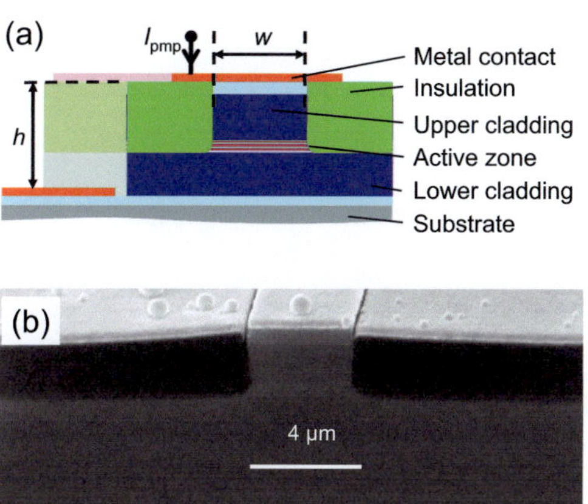

Fig. 8.2. (a) Waveguide cross section of the quantum dot semiconductor optical amplifier ($w = 4\,\mu$m, $h \approx 2\,\mu$m). (b) SEM picture of the waveguide facet. Pictures courtesy of M. Lämmlin [61].

8.3 Device Model

In the pump-probe experiment, the amplitude and the phase of the probe impulse are measured as a function of the pump-probe delay τ, see Chapter 2 for a detailed description. For sufficiently large positive values of τ, the pump and the probe impulses do not overlap. In this case, coherent nonlinear interaction between the pump and the probe signal can be neglected, and the pump-induced change of the probe amplitude and phase can be described in terms of a refractive index change Δn and a material power gain change Δg in the active region of the waveguide. Inside the active region, the complex refractive index profile thus changes by $\Delta n + \mathrm{j}\,\Delta g/(2k_0)$. Since the pump impulse may gain or loose energy during propagation along the active waveguide, Δn and Δg depend both on the longitudinal coordinate z and on the pump-probe delay τ.

According to Eqs. (1.57) and (1.60) the evolution of the probe impulse during propagation along the waveguide is governed by

$$\frac{\partial A_{\mathrm{prb}}(z,t)}{\partial z} = -\mathrm{j}\,k_0\Gamma\left(\Delta n(z,\tau) + \mathrm{j}\,\frac{\Delta g(z,\tau)}{2k_0}\right) A_{\mathrm{prb}}(z,t), \qquad (8.1)$$

where $k_0 = 2\pi/\lambda$ is the vacuum wavenumber and where Γ denotes the field confinement factor for the active region, see Eq. (1.61). Since the active waveguide is very short, deformation of the envelope due to dispersion can be neglected. A retarded time frame that travels along with the group velocity of the impulse has been introduced already, see Eq. (1.54). Solving Eq. (8.1), the amplitude transmission factor $T(\tau)$ and the phase change $\Delta\phi(\tau)$ as defined in Eq. (2.26) are obtained

$$T(\tau) = e^{\frac{\Gamma}{2}\int_0^L \Delta g(z,\tau)\,\mathrm{d}z} \,, \tag{8.2}$$

$$\Delta\phi(\tau) = -\Gamma \int_0^L k_0 \Delta n(z,\tau)\,\mathrm{d}z, \tag{8.3}$$

To a first-order approximation, the gain change $\Delta g(z,\tau)$ and index change $\Delta n(z,\tau)$ are proportional to the pump-induced deviation of different carrier populations from their steady-state values. After the pump-impulse has passed, the different nonequilibrium carrier concentrations show an exponential decay. Therefore, the material gain perturbation $\Delta g(z,\tau)$ and the refractive index perturbation $\Delta n(z,\tau)$ can be assumed to exhibit exponentially decaying behaviour with respect to the pump-probe delay τ. We assumed a superposition of two exponential decays with independent time constants τ_1, τ_2 for the material gain and τ_a, τ_b for the refractive index,

$$\Delta g(z,\tau) = \Delta g_1(z)\,e^{-\frac{\tau}{\tau_1}} + \Delta g_2(z)\,e^{-\frac{\tau}{\tau_2}} \,, \tag{8.4}$$

$$\Delta n(z,\tau) = \Delta n_a(z)\,e^{-\frac{\tau}{\tau_a}} + \Delta n_b(z)\,e^{-\frac{\tau}{\tau_b}} \,. \tag{8.5}$$

The measured probe transmission $T_{\mathrm{dB}}(\tau)$ in dB and the phase change $\Delta\phi(\tau)$ thus decay also exponentially with time constants τ_1, τ_2 and τ_a, τ_b,

$$T_{\mathrm{dB}}(\tau) = 20\log_{10}\left(T(\tau)\right) = T_{\mathrm{dB},1}\,e^{-\frac{\tau}{\tau_1}} + T_{\mathrm{dB},2}\,e^{-\frac{\tau}{\tau_2}} \,, \tag{8.6}$$

$$\Delta\phi(\tau) = \Delta\phi_a\,e^{-\frac{\tau}{\tau_a}} + \Delta\phi_b\,e^{-\frac{\tau}{\tau_b}} \,, \tag{8.7}$$

where

$$T_{\mathrm{dB},1} = 4.343\,\Gamma \int_0^L \Delta g_1(z)\,\mathrm{d}z, \quad T_{\mathrm{dB},2} = 4.343\,\Gamma \int_0^L \Delta g_2(z)\,\mathrm{d}z, \tag{8.8}$$

and

$$\Delta\phi_a = -k_0\Gamma \int_0^L \Delta n_a(z)\,\mathrm{d}z, \quad \Delta\phi_b = -k_0\Gamma \int_0^L \Delta n_b(z)\,\mathrm{d}z. \tag{8.9}$$

8.4 Measurement Results and Interpretation

We used the heterodyne pump-probe setup described in Section 2.5 to investigate the dynamics of the quantum-dot SOA. The center wavelength of the pump and probe impulses was tuned to $\lambda = 1\,299\,\mathrm{nm}$, which is close to the gain maximum of the SOA, and the complex amplitude transmission factor $T(\tau)\,e^{\mathrm{j}\,\Delta\phi(\tau)}$ according to Eq. (2.26) was measured as a function of the pump-probe delay τ. The corresponding traces for different pump impulse energies measured in the input fiber are shown in Fig. 8.3. The pump current was held constant at $100\,\mathrm{mA}$ (current density: $1.25\,\mathrm{kA}/\mathrm{cm}^2$) for all cases.

The amplitude and phase parameters $\tau_1, \tau_2, T_{\mathrm{dB},1}, T_{\mathrm{dB},2}$, and $\tau_a, \tau_b, \Delta\phi_a, \Delta\phi_b$ were extracted from fitting the measured data. For both amplitude and phase, we found a

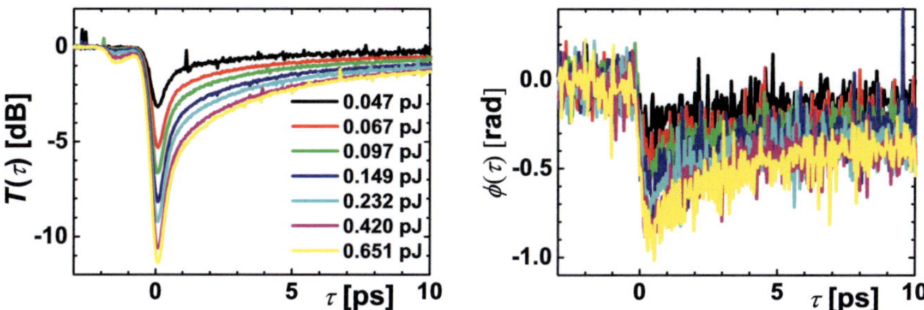

Fig. 8.3. Amplitude and phase response of a quantum-dot semiconductor optical amplifier, operated at a wavelength of $\lambda = 1300\,\mathrm{nm}$.

quickly decaying process with nearly equal time constants $\tau_1 \approx 3.00 \pm 0.24\,\mathrm{ps}$ and $\tau_a \approx 2.63 \pm 0.37\,\mathrm{ps}$. The time constants τ_2 and τ_b were both of the order of several hundred picoseconds. We used different pump impulse energies as indicated in Fig. 8.3 at a constant SOA pump current of $100\,\mathrm{mA}$. Within the measurement accuracy, the time constants did not show any systematic dependence on the impulse energy. On the other hand, the small time constants are found to decrease when increasing the pump current at a constant pump impulse energy of $W_{\mathrm{pmp}} = 0.13\,\mathrm{pJ}$. We attribute the small time constants to carrier capture from the wetting layer into the QD states, and the large ones to the recovery time of the wetting layer.

For each of these processes, the pump-induced local change in material gain is approximately proportional to the corresponding refractive index change, where the proportionality factor is related to the α-factor (Henry factor) α_{H} [48]. From the relative magnitudes of the gain and the phase changes the α-factor α_{H1} for the fast carrier capture process and for the slow α_{H2} recovery process may be derived,

$$\alpha_{\mathrm{H1}} = -\frac{4\pi}{\lambda}\frac{\Delta n_a}{\Delta g_1} = 8.686\frac{\Delta\phi_a}{T_{\mathrm{dB},1}} \approx 0.92 \pm 0.22, \tag{8.10}$$

$$\alpha_{\mathrm{H2}} = -\frac{4\pi}{\lambda}\frac{\Delta n_b}{\Delta g_2} = 8.686\frac{\Delta\phi_b}{T_{\mathrm{dB},2}} \approx 3.22 \pm 0.54. \tag{8.11}$$

The fast process has a remarkably small α-factor. This indicates that the fast nonlinear processes mainly affect the gain rather than the phase. The α-factor of the slower carrier recovery process is more than three times bigger. It thus must be concluded that fast signal processing schemes based QD SOA should exploit cross-gain modulation and, to avoid pattern effects, at the same time be insensitive to cross-phase modulation.

8.5 Summary

Strong 3 ps gain variations with only weak phase changes were measured with a pump-probe setup on an InAs/GaAs quantum dot amplifier at $1.3\,\mu\mathrm{m}$. Such low-alpha factor devices are well suited for cross-gain modulation based signal processing.

Chapter 9

Resonant Enhancement of Nonlinear Interaction: Ring Resonators

9.1 Introduction

Nonlinear resonators are of considerable technical interest, since they can be used for ultra-fast all-optical signal processing at low power levels, e.g., for the realization of logic operations [76] or fully transparent wavelength conversion [1]. Industrial and research design of such novel optical components heavily depends on the reliability and the accuracy of mathematical modelling algorithms. A very general tool for simulating integrated optical devices is the finite-difference time-domain (FDTD) method [123, 106].

FDTD directly discretizes Maxwell's equations in the time domain. It is thus a robust method that contains a minimum amount of implicit assumptions. The FDTD method has been previously applied to various nonlinear optical problems. Optical soliton propagation has been investigated [40, 106], and quantum effects in optical gain media have been described by solving Maxwell-Bloch equations and electron population rate equations [96, 17]. A linear two-dimensional (2D) FDTD analysis of microcavity optical ring and disk resonators can be found in [106], and a nonlinear 2D FDTD method for simulating second and third-order nonlinear effects in nondispersive photonic crystal (PhC) structures is reported in [87].

However, due to the prohibitive computational expenditure, the FDTD method has been neither applied to realistic three-dimensional (3D) nonlinear dispersive resonators, nor has its ability of modelling such structures been confirmed by experiments. A fully 3D simulation of all effects makes the algorithm computationally expensive and is therefore not viable for many practical applications. For using FDTD algorithms for design purpose rather than as verifications tools, it is therefore essential to know under which circumstances a simpler model suffices.

The scope of this chapter is twofold: First, we present the results obtained from an advanced FDTD algorithm generalized to incorporate all physical effects that are necessary for modelling a realistic case of nonlinear interaction in a dispersive resonator. As an example of practical interest, we have chosen parametric four-wave-mixing (FWM) in an integrated InGaAsP/InP racetrack resonator of realistic size. The algorithm is fully 3D and takes into account $\chi^{(3)}$-nonlinearities and material dispersion. It further compensates for staircasing errors by using the effective dielectric constant (EDC) technique [32]. This is, to the best of our knowledge, the most realistic FDTD modelling of nonlinear interaction in a dispersive resonator structure published so far. In order to provide the considerable numerical resources, the code was implemented on a parallel computer. We have fabricated and characterized the simulated structure and can thus investigate the reliability of the implementation. An analytical treatment of the resonant nonlinear interaction allows us to attribute the discrepancies between the simulation and the experiment to geometrical fabrication imperfections, whereas the method itself is found to be accurate.

Second, having confirmed the reliability of the complex model, we use its results as a benchmark to evaluate a series of progressively less complicated, but computationally more manageable models. In several steps, we replace the 3D algorithm by a 2D effective index approximation, neglect material dispersion, and leave out the EDC smoothing. The influences of these simplifications on the free spectral range (FSR) of the resonator and the FWM conversion efficiency are investigated, and we can finally give guidelines on how realistic a model must be in order to predict the device's behaviour to a given degree of accuracy.

The chapter is structured as follows: In Section 2 we give an analytical expression for the resonant FWM-efficiency. In Section 3 we describe the device structure and the numerical procedure. In Section 4 we outline the fabrication of the test structure as well as the details of the conversion efficiency measurement. In Section 5 we finally compare the results obtained from the different FDTD implementations and give guidelines as to which kind of implementation is necessary to model different aspects of the nonlinear resonator.

Parts of the following chapter have been published in journal articles [J1, J3], see pp. 199.

9.2   Theory of FWM in a Ring Resonator

In order to understand the physics of FWM in resonant structures it is necessary to have an analytical model which incorporates the main physical phenomena. For this reason we start with a simple transmission-line analysis of FWM in high-contrast ring-resonators. This model allows us to explicitly formulate the connection between the field enhancement and the conversion efficiency. We use an expression for the FWM conversion efficiency which takes into account the full vectorial mode fields of high index-contrast waveguides.

The device under investigation is depicted in Fig. (9.1): It consists of a racetrack-shaped ring resonator which is coupled to two straight bus waveguides. The incoming and the outgoing wave amplitudes are denoted as a_n and b_n, respectively, where n is the port index and the waves inside the ring are labelled with an additional subscript "i". The amplitude coupling factors of the coupling zones are denoted as $\kappa_{1,2}$, and the transmission factors are $\mu_{1,2}$. We assume the coupling zones to be perfectly matched and lossless, i.e., $\mu_{1,2}^2 + \kappa_{1,2}^2 = 1$. Propagation losses inside the ring are taken into account by a round-trip amplitude transmission factor $\rho \in [0, 1]$.

If the structure is excited with a signal close to a resonance frequency, a strong wave builds up in the ring, and the field inside the ring is enhanced by a factor of FE $= |b_{2\mathrm{i}}/a_1|$

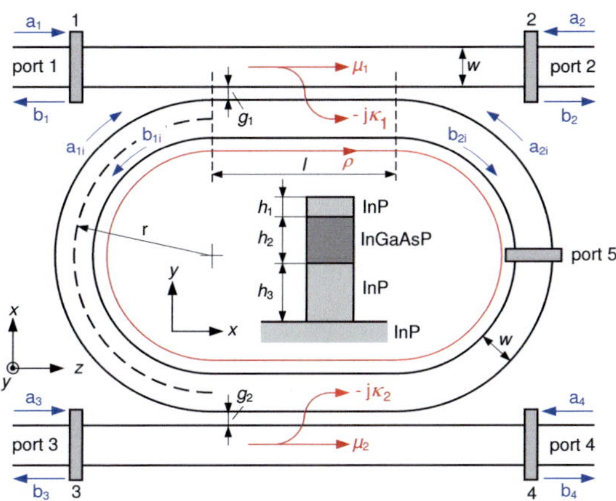

Fig. 9.1. Configuration of the racetrack resonator. The inset shows the cross section of the pedestal waveguide. The dimensions are as follows [μm]: $w = 0.4$, $l = 3$, $r = 5$, $g_1 = 0.15$, $g_2 = 0.2$, $h_1 = 0.3$, $h_2 = 0.7$, $h_3 = 1.5$.

with respect to the exciting wave in the bus waveguide. FE is referred to as the field enhancement factor, and can be approximated by

$$\mathrm{FE} = \mathrm{FE}_{\mathrm{res}} \frac{1}{\sqrt{1 + \left(2\frac{\Delta\omega}{\Delta\omega_{\mathrm{FWHM}}}\right)^2}}, \tag{9.1}$$

in the vicinity of a resonance line. Here $\mathrm{FE}_{\mathrm{res}} = \kappa_1 / (1 - \rho\mu_1\mu_2)$ is the resonant field enhancement, $\Delta\omega$ denotes the detuning from the nearest resonance line and $\Delta\omega_{\mathrm{FWHM}}$ is the corresponding full-width at half maximum (FWHM).

In degenerate FWM processes, the nonlinear interaction of two waves at frequency ω_s ("signal") and ω_p ("pump") leads to the formation of a third wave at frequency $\omega_c = 2\omega_p - \omega_s$ ("converted"). The conversion efficiency η_{FWM} is the ratio of the incoming signal power divided by the outgoing converted power. This ratio can be improved considerably by exploiting the field enhancement [1]. In Section 1.5.3 we have derived the equations for the conversion efficiency in a straight high index-contrast (HIC) waveguide, taking into account the vectorial electric field $\mathcal{E}(x, y, \omega_\nu)$ and magnetic field $\mathcal{H}(x, y, \omega_\nu)$ of the waveguide mode. Taking into account the resonant field enhancement, this analysis yields relations identical to the ones given in [1] and in the references therein, but with a different expression for the nonlinear waveguide parameter γ. In particular, we obtain

$$\eta_{\mathrm{FWM}} = \mathrm{FE}_p^4 \, \mathrm{FE}_s^2 \, \mathrm{FE}_c^2 \left|\gamma^2 P_p^2 L_{\mathrm{rt,eff}}\right|^2, \tag{9.2}$$

where P_p is the pump power and $\mathrm{FE}_{p,s,c}$ are the field enhancement factors of the pump, signal and converted waves, respectively. $L_{\mathrm{rt,eff}}$ is the resonator's effective round-trip length for nonlinear interaction and takes into account losses in the ring waveguide, phase-mismatch due to second-order waveguide dispersion, and losses due to coupling to a second bus waveguide. Following Eq. (1.78), the nonlinear waveguide parameter γ now depends on the vectorial electric and magnetic mode fields,

$$\gamma(\omega_{\mathrm{c}} : \omega_{\mathrm{p}}, \omega_{\mathrm{p}}, -\omega_{\mathrm{s}}) = \frac{3\omega_c \epsilon_0}{16\sqrt{\mathcal{P}(\omega_{\mathrm{c}})\mathcal{P}(\omega_{\mathrm{p}})\mathcal{P}(\omega_{\mathrm{p}})\mathcal{P}(\omega_{\mathrm{s}})}}$$
$$\times \iint \left[\chi^{(3)} : \mathcal{E}^*(\omega_{\mathrm{s}})\mathcal{E}(\omega_{\mathrm{p}})\mathcal{E}(\omega_{\mathrm{p}}) \right] \cdot \mathcal{E}^*(\omega_{\mathrm{c}})\, \mathrm{d}x\, \mathrm{d}y. \qquad (9.3)$$

$\chi^{(3)}$ is the third-order nonlinear susceptibility of the waveguide material, and the quantities $\mathcal{P}(\omega)$ are used for power-normalization of the modal fields, see Eq. (1.40),

$$\mathcal{P}(\omega_\nu) = \frac{1}{2} \iint \mathrm{Re}\left\{ (\mathcal{E}(x,y,\omega_\nu) \times \mathcal{H}^*(x,y,\omega_\nu)) \right\} \cdot \mathbf{e}_z\, \mathrm{d}x\, \mathrm{d}y, \qquad (9.4)$$

where $\mathbf{e}_z$ is a unit vector pointing along the local direction of the waveguide axis.

We note that the underlying transmission line model of the ring resonator cannot be used directly to make predictions because the field enhancement in the resonator relies on the very 3D structure of the device. An additional calculation, e.g., based on one of the more elaborate FDTD algorithms described below, is necessary to calculate FE.

9.3   FDTD Simulation of a Racetrack Micro-Resonator

9.3.1   Reference Structure

The nonlinear racetrack micro-resonator of Fig. 9.1 is used as a reference structure for evaluating the various FDTD implementations. The waveguides consist of a Q(1.35) InGaAsP core sandwiched between an InP pedestal and cap. The refractive indices of both materials are affected by material dispersion, which was taken into account where indicated. Otherwise, the refractive indices were assumed to take the constant values of $n = 3.42$ for InGaAsP and $n = 3.17$ for InP at a wavelength of 1.55μm. The structure is surrounded by air ($n = 1$).

9.3.2   Simulation Method

As is common for all FDTD algorithms, Maxwell's curl equations

$$\nabla \times \mathbf{H} = \frac{\partial \mathbf{D}}{\partial t}, \quad \nabla \times \mathbf{E} = -\mu \frac{\partial \mathbf{H}}{\partial t}, \qquad (9.5)$$

are solved using Yee's leapfrog algorithm [123, 33, 106]. Nonlinearity and material dispersion are introduced via the constitutive equation for isotropic media

$$D_\xi = \varepsilon_0 \varepsilon_\infty E_\xi + \varepsilon_0 \chi^{(1)} E_\xi + \varepsilon_0 \chi^{(3)} |\mathbf{E}|^2 E_\xi , \qquad (9.6)$$

where ξ stands for a Cartesian field component (x, y, or z), D_ξ denotes the component of the electric displacement vector, and ε_∞ the relative dielectric constant in the limit of infinite frequency. The frequency-dependent susceptibility $\chi^{(1)}$ determines the material dispersion, whereas the nonlinear coefficient $\chi^{(3)}$ was assumed to be frequency-independent and has been previously measured to be $\chi^{(3)} = 3.8 \times 10^{-18}\,\mathrm{m^2/V^2}$ for an InGaAsP alloy similar to ours [21]. This value of $\chi^{(3)}$ was used in all simulations. We assumed that the nonlinear and dispersive effects exist only for the InGaAsP core region of the waveguide, because the field is strongly confined in the guiding region and the nonlinearity is stronger in InGaAsP than in InP within the frequency range of interest. The dielectric loss of the media and possible losses associated with the roughness of the waveguide walls were neglected in the FDTD analysis.

The material dispersion of InGaAsP, precisely modelled by Adachi [2], can be approximated using a Lorentz model in the frequency domain,

$$\chi^{(1)} = \frac{\Delta\epsilon\,\omega_p^2}{\omega_p^2 - \omega^2} . \tag{9.7}$$

In the frequency range of interest, the Lorentz model gives an excellent approximation with a maximum relative error of 2×10^{-4}. In the time domain, the dispersion can be represented using a differential equation of second order. The parameters $\epsilon_\infty = 10.05$, $\Delta\epsilon = 1.0$ and $\omega_p = 1.948 \times 10^{15}$ rad/s were calculated to fit the material dispersion for the frequency range from 150 THz to 210 THz. The full details of our numerical treatment of nonlinearities and dispersion can be found in [33] [31] [30].

A persistent difficulty in FDTD methods is the staircasing error, i.e., an error that is caused by the inaccurate representation of a curved dielectric surface using a grid in Cartesian coordinates. Even for a very fine spatial discretization, staircasing can lead to numerical scattering and back-reflection which do not exist in the physical device. This error can be mitigated by applying the effective dielectric constant (EDC) technique, which homogenizes the curved surfaces within a Yee cell by replacing the dielectric constant of the affected Yee cells with an anisotropic effective dielectric constant calculated at long wavelengths [32].

9.3.3 Simulation Parameters

For all simulations, the computational area in the (x, z) plane was chosen to be $13.5 \times 17.4\ \mu\mathrm{m}^2$, with the racetrack in the center of the simulation region. For the 3D analysis, we included a 0.2 μm thick part of the substrate, and a 0.5 μm thick part of the air region above the waveguide. In this case, all boundaries of the calculation region are set to be perfectly matched layers (PML). As an excitation field, we launched the TM-polarized (dominant component of the electric field perpendicular to the substrate plane) fundamental waveguide mode into port 1.

For the 2D models, the standard 3D algorithm was used, but the calculation region in the y direction was chosen to be only a single Yee cell in height. Perfect electric conductor (PEC) boundary conditions were then applied on the upper and lower planes in the y-direction, and the effective index method (EIM) was used to transform the 3D structure into a 2D effective index profile [19]. This results in an effective refractive index $n_\mathrm{2D} = 3.34$ for the InP/InGaAsP/InP layer stack. For the 2D model that includes material dispersion, the dispersion curve given by Eq. (9.7) was adapted accordingly.

The spatial discretization was chosen to be $\Delta z = \Delta x = 0.025\ \mu$m, which approximately corresponds to $1/20$ of the guided wavelength, with $\Delta y = 0.1\ \mu$m in the 3D case. The time step Δt was 0.0465 fs, equivalent to 0.8 times the maximum allowable Courant limit $\Delta t_c = \left(\Delta x^{-2} + \Delta y^{-2} + \Delta z^{-2}\right)^{-0.5}/c$. The computations were performed on a parallel cluster (HP XC6000) at the University of Karlsruhe. Using 64 processors, the computation time was approximately $1\,700$ minutes for the complete 3D simulation of the reference structure including both material dispersion and EDC smoothing.

The resonance frequencies of the ring were first calculated by launching a low-power band-limited short pulse with a center frequency of 192 THz and a bandwidth of 12 THz from port 1. The calculation was performed for the duration of 10 ps. The frequency spectra were then obtained by a Fourier transform of the time signal.

In order to calculate the nonlinear conversion efficiency, we chose the frequencies for the pump and the signal waves such that they matched exactly the resonance frequencies obtained from the linear simulation. Continuous sinusoidal waves of pump and signal were both input from port 1 of the bus waveguide. The intensity of the signal wave was chosen to be smaller by 20 dB than that of the pump wave.

The excitation for the 2D simulations was chosen such that the maximum electric field amplitude in the center of the waveguide was the same as in the 3D case. The time varying electric fields were detected for a duration of 18 ps at the center of each port depicted in Fig. 9.1. Fourier spectra were then computed from the time signals to find the wavelength conversion efficiency and the field enhancement factors. In order to detect the converted wave, which has a very small amplitude, a raised-cosine time window (Hann time window) was used when applying the Fourier transform. An example of the calculated spectrum is shown in Fig. 9.2. The field enhancement factor and the wavelength conversion efficiency can be calculated from the ratios of the corresponding peaks. Note that a number of peaks other than the pump and signal waves are observed — these are the results of subsequent nonlinear interactions among various lines. The width of the spectral lines in Fig. 9.2 is caused by the limited duration of the simulated time span.

9.4   Fabrication and Measurement

To fabricate the previously discussed racetrack resonator, the appropriate layer stack was grown on top of an n-doped InP wafer, and a thin silicon-nitride (SiN_x) layer was deposited using plasma-enhanced chemical vapor deposition (PECVD). Then the wafer was spin-coated with a negative-tone electron beam resist, and the proximity corrected pattern was exposed using direct write electron beam lithography. After development and hard-bake, the SiN_x-mask was patterned by means of reactive ion etching (RIE), and finally the waveguides were created by a chlorine-based chemically-assisted ion beam etching (CAIBE) process. The samples were cleaved into bars and mounted using standard techniques.

For the measurement, polarization-maintaining lensed fibers were used to couple TM-polarized light to the bus waveguide of the device. First, a characterization of the linear transmission properties of the resonator was performed at low power. To obtain the resonator parameters κ_n, μ_n (subscript $n = 1$ for the input, $n = 2$ for the output coupling

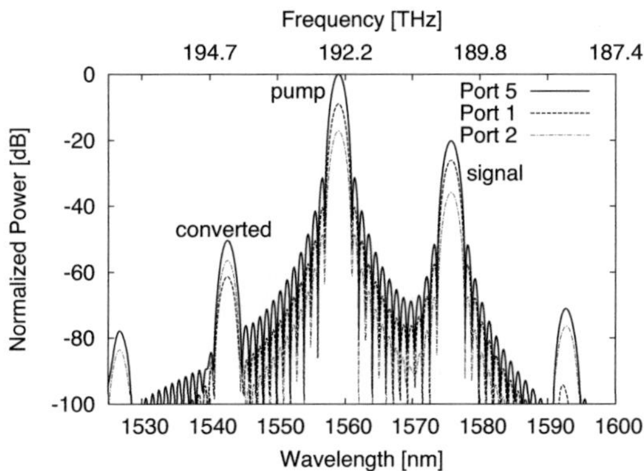

Fig. 9.2. Normalized power spectrum at different ports for a maximum pump electric field $E_p = 5.65 \times 10^6$ V/m ($P_p \approx 13\,\text{dBm}$) obtained from the 3D FDTD analysis including material dispersion and omitting EDC. E_p is the dominant component of the pump electric field measured in the center of the waveguide, and P_p is the corresponding pump power.

zone) and ρ as defined in Fig. 9.1, the measured results were fitted by a transmission-line model of the ring resonator, where we assume lossless coupling zones ($\kappa_n^2 + \mu_n^2 = 1$). For a preliminary fit, only the coupling coefficients κ_n, the round-trip amplitude transmission ρ and the round-trip group delay in the center of the measured spectral range were used as fit parameters. The results served as initial values for a subsequent fit, in which a linear frequency-dependence of all parameters was assumed and the associated slopes with respect to frequency were also used as fit parameters. The measured curve and the fitted curve exhibit very good agreement, as shown in Fig. 9.3. In the center of the measured frequency range, we found coupling coefficients κ_n and transmission coefficients μ_n for the input and the output coupling zones of $\kappa_1 = 0.244$, $\mu_1 = \sqrt{1 - \kappa_1^2} = 0.970$, $\kappa_2 = 0.127$, $\mu_2 = \sqrt{1 - \kappa_2^2} = 0.992$. The round-trip amplitude transmission of the resonator was found to be $\rho = 0.922$. Using these values, a resonant field enhancement of $\text{FE}_{\text{res}} = 2.16$ is extracted from the measurement.

For the investigation of nonlinear optical interaction in the resonator, a high-power pump wave (>10 dBm on chip) was launched at the resonance near the frequency $f_2 = 192.416\,\text{THz}$ (wavelength $\lambda_2 = 1\,558.044\,\text{nm}$), the signal was tuned to the resonance near $f_1 = 190.395\,\text{THz}$ ($\lambda_1 = 1\,574.582\,\text{nm}$), and the converted wave was observed at $f_3 = 2f_2 - f_1$. The strong pump leads to heating of the ring, and the resonance comb shifts to higher frequencies. Therefore, for each pump power, the pump and signal frequencies had to be adjusted to track the shifted resonances.

Figure 9.4 shows an example for the spectrum measured at port 2. One can clearly see the pump, signal, and converted lines. In order to calculate the conversion efficiency, the on-chip power of the outgoing converted wave must be divided by the on-chip power of the incoming signal, and so the coupling loss between the fiber and the chip, as well as the

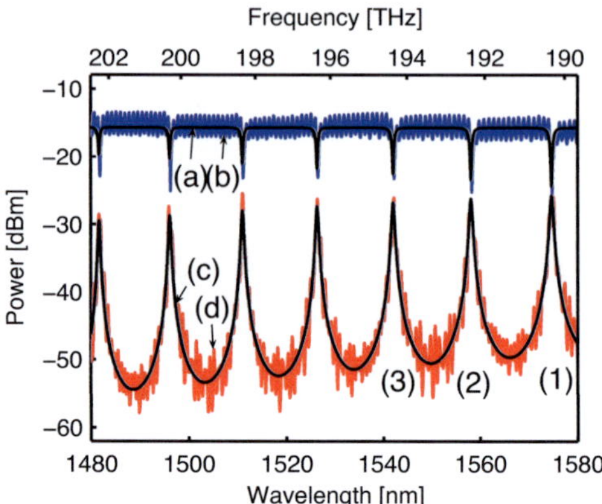

Fig. 9.3. Measured transmission spectra and fitting results for the bus port 2, fitted curve (a) and measured (noisy) curve (b), and for the drop port 3, fitted curve (c) and measured (noisy) curve (d). The numbers (1), (2) and (3) denote the resonances that are used for resonant FWM and for comparison with the simulation.

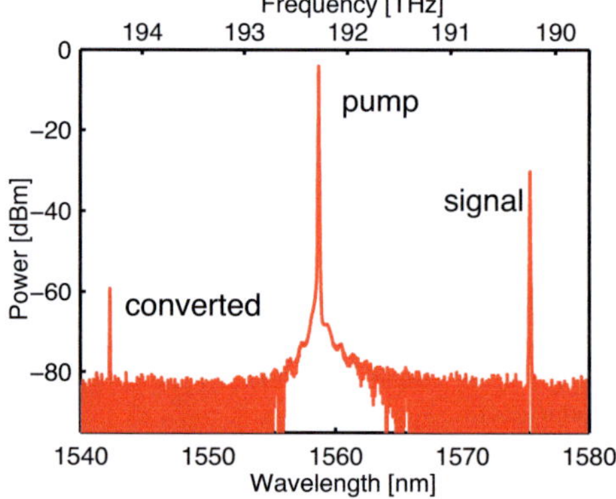

Fig. 9.4. Measured spectrum at the output (bus port 2) for FWM.

propagation loss in the bus waveguide, must be taken into consideration. This was done by measuring the transmission of straight reference waveguides on the same bar, which were not coupled to a ring resonator. The highest conversion efficiency thus obtained amounts to −32.6 dB (see Fig. 9.7). This is, to the best of our knowledge, the highest conversion efficiency in a passive micro-resonator that has been reported so far. It can be further increased by using higher pump powers, but will eventually be limited by two-photon absorption, which generates free carriers and thus induces free-carrier absorption.

9.5 Verification and Comparison of the FDTD Results

To cover all physical effects that are relevant for resonant FWM, our most advanced FDTD solver simultaneously incorporates χ^3-nonlinearities and material dispersion, and it corrects staircasing errors. This implementation is used as a benchmark for evaluating the reliability of the simplified algorithms, which are obtained by ignoring material dispersion, by omitting the staircasing error correction, by using a 2D effective index method [19], or by combining any of these simplifications. The different algorithms were applied to the reference structure described in the previous sections, and simulations were performed in the linear and nonlinear regime of the resonator.

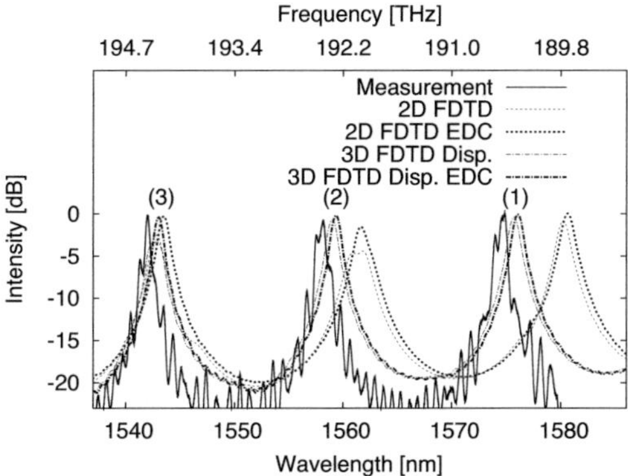

Fig. 9.5. Normalized transmission spectra detected at the drop port (port 3) obtained by measurement and FDTD analyses with and without using EDC smoothing. The measured curve appears noisy due to multiple reflections from the waveguide facets (device facets were not anti–reflection coated). The numerals (1), (2), and (3) in the figure indicate the resonances from which the frequencies f_1, f_2, and f_3 (Table 9.5) were extracted. The resonances (1) and (2) are chosen for launching the signal and the pump waves for the investigation of FWM, and the converted wave emerges in the vicinity of resonance (3).

Figure 9.5 shows the normalized drop port (port 3) transmission spectra for the measurement and various FDTD analyses. From each linear transmission analysis, we have extracted three resonance frequencies f_1, f_2, f_3 in the vicinity of $1\,550\,\mathrm{nm}$ indicated by numbers (1), (2), (3) in Fig. 9.5. These frequencies and the corresponding free spectral ranges $\Delta f_{3,2} = f_3 - f_2$ and $\Delta f_{2,1} = f_2 - f_1$ are listed in Table 9.5. Further, we calculated the deviation of each value from its measured counterpart and normalized the result by the measured mean free spectral range $\Delta f^{(\mathrm{meas})} = (\Delta f_{3,2}^{(\mathrm{meas})} + \Delta f_{2,1}^{(\mathrm{meas})})/2 = 2.016\,\mathrm{THz}$. The

Method		f_1	$\Delta f_{2,1}$	f_2	$\Delta f_{3,2}$	f_3
measurement		190.395	2.021	192.416	2.010	194.426
3D	disp., EDC	190.220	2.043	192.263	2.034	194.297
	disp. only	190.262	2.044	192.306	2.032	194.338
	EDC only	189.933	2.191	192.124	2.192	194.316
	none	189.948	2.228	192.176	2.191	194.367
2D	disp., EDC	190.106	2.097	192.203	2.083	194.286
	disp. only	190.147	2.102	192.249	2.095	194.344
	EDC only	189.675	2.287	191.962	2.274	194.236
	none	189.725	2.287	192.012	2.302	194.314

Table 9.1. Resonance frequencies f_1, f_2, and f_3 and corresponding local free spectral ranges $\Delta f_{2,1} = f_2 - f_1$ and $\Delta f_{3,2} = f_3 - f_2$ in THz.

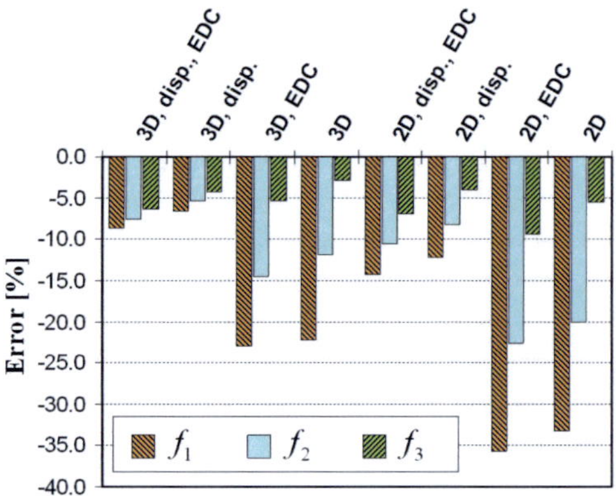

Fig. 9.6. Comparison of measured and calculated resonance frequencies for the different FDTD implementations. The bars indicate the deviation of the resonance frequencies f_1, f_2, and f_3 in percent of the averaged measured FSR $\Delta f^{(\mathrm{meas})} = 2.016\,\mathrm{THz}$.

resulting error bars are plotted in Fig. 9.6. The differences in length between neigbouring bars represent the relative errors of the corresponding FSR.

For the nonlinear simulations, the FWM conversion efficiencies η_{FWM} were calculated using the ratio of the power of the outgoing converted wave P_c and that of the incoming signal wave P_s, i.e., $\eta_{\mathrm{FWM}} = |P_c|/|P_s|$. The results are plotted in Figs. 9.7, 9.10, and will be discussed in the following section. For each FDTD model, we simulated the conversion efficiency for three different pump powers. This is sufficient since we know from Eq. (9.2) that the data points in the logarithmic plot have to lie on a straight line with a slope of $20\,\mathrm{dB}$ per decade (in terms of pump power).

9.5.1 Measurement and Benchmark Simulation

We first compare the measurement with the benchmark FDTD simulation, which includes χ^3-nonlinearities, material dispersion, and correction for staircasing errors. As can be seen from Table 9.5 and Fig. 9.6, the computed spectrum is down-shifted in frequency with respect to the measurement. The differences between the measured and the simulated resonance frequencies amount to roughly 8 % of the measured FSR (which corresponds to less than 0.1 % of the respective resonance frequencies). The displacement of the resonances is quite uniform over the spectral range of interest, resulting in a simulated FSR which deviates from the measured one by only 1 %. We may thus conclude that the simulated dispersion relation of the ring waveguide differs from the measurement mainly by an overall shift, whereas the slope is essentially the same.

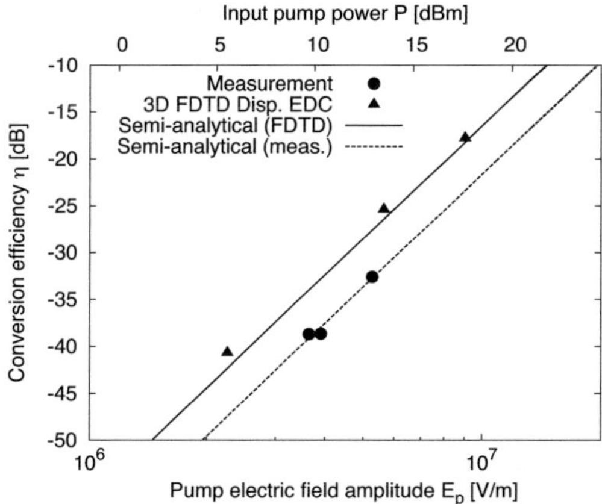

Fig. 9.7. Comparison of the conversion efficiencies obtained by the measurement, by the 3D FDTD analysis including the material dispersion and EDC (benchmark model), and by the semi-analytical formulae based on the respective field enhancement factors.

In Fig. 9.7, the conversion efficiency η_{FWM} is displayed in a double-logarithmic plot versus the peak electric field E_p of the pump wave (lower axis) or the corresponding pump power P_p (upper axis). The measured values (●) and the simulated efficiencies (▲) lie on two lines with identical slope. The simulated conversion efficiencies are 5 dB larger than the measured values.

To isolate the reason for this difference, we compare the measurement and the simulation with the predictions of Eq. (9.2), indicated as straight lines in Fig. 9.7. The most sensitive terms in Eq. (9.2) are the nonlinearity parameter γ and the field enhancement factors $\mathrm{FE}_{\mathrm{p,s,c}}$. The nonlinearity parameter were computed by evaluating the overlap integral according to Eq. (9.3), using numerically calculated fully-vectorial mode fields of the straight waveguides. The field enhancement factors of the pump-, signal- and converted wave were assumed to be identical, $\mathrm{FE}_{\mathrm{p,s,c}} = \mathrm{FE}$. For the upper line, the field enhancement factor $FE_{calc} = 2.40$ was obtained by evaluating the FDTD data directly, and the effective round-trip length was taken to be the geometrical length $|L_{rt,eff,calc}| = L_{rt,geom}$ corresponding to a round-trip amplitude transmission factor $\rho_{calc} = 1$. The lower line (Semi-analytical (meas)) is based on the field enhancement $FE_{meas} = 2.16$ and the round-trip amplitude transmission $\rho_{meas} = 0.922$ that were obtained from fitting the measured linear transmission spectrum as described in Section 9.4. In this case, the effective round-trip length has to be reduced accordingly, $|L_{rt,eff,meas}| \approx \rho_{meas}^2 \times L_{rt,geom}$.

For both the measurement and the benchmark simulation, the agreement with the analytical model according to Eq. (9.2) is excellent, and it can be concluded that the discrepancy between the measured and the simulated efficiency is only due to different field enhancement factors and different losses in the ring waveguide. We may further infer that the semi-analytical model gives a reliable prediction of the conversion efficiency once the field enhancement is known. Also, the agreement shows that the value for $\chi^{(3)}$ used in the simulations is correct.

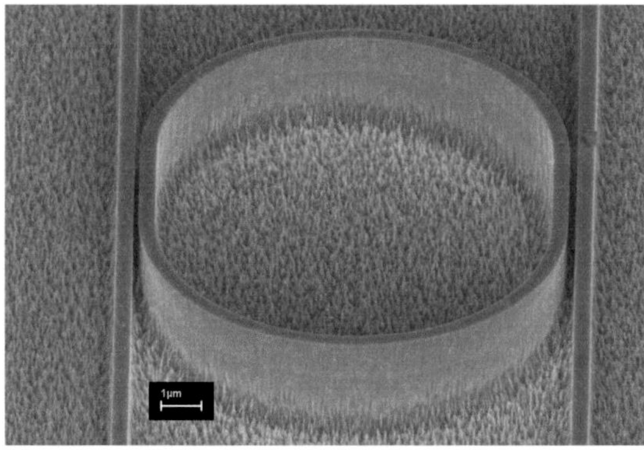

Fig. 9.8. Scanning electron microscope (SEM) picture of the fabricated racetrack micro-resonator.

This leads to the question why the dispersion relation is shifted, and why the field enhancement factors and the resonator losses in the simulation and in the measurement are so different. In the mask layout, the waveguide width was designed to be 400 nm, and the coupling gaps between the bus waveguides and the resonator were chosen to be $g_1 = 0.15\,\mu m$ for the input side and $g_2 = 0.20\,\mu m$ for the output side. These values were used in the FDTD simulation. However, by dissecting the structure using focussed ion beam milling (FIB), we found that the width of the fabricated pedestal waveguides was slightly smaller than the one used in the simulation. This was caused by lithography tolerances and a small etch undercut. An actually reduced waveguide width decreases the effective refractive index of the fundamental mode and shifts the measured resonance lines to higher frequencies, which is consistent to our observation.

Further, the high aspect-ratio coupling gaps were not completely etched down to the substrate, and were wider than designed due to the undercut of the waveguide (actual values $g'_1 \approx 0.185$ μm and $g'_2 \approx 0.240$ μm in the center of the waveguide core). This affects the coupling factors $\kappa_{1,2}$. It can also be seen from the SEM picture Fig. 9.8 that the sidewalls of the waveguides have a certain roughness. This leads to scattering losses that were not taken into account in the simulation. However, the field enhancement FE depends on both the coupling factors $\kappa_{1,2}$ and the round-trip attenuation factor ρ, see Eq. (9.1). Therefore, FE reacts very sensitively to surface roughness and small geometrical fluctuations in the coupling zone. The conversion efficiency is proportional to the 8th power of the field enhancement and will thus depend even more sensitively on the small-scale geometry variation of the device.

Random deviations such as surface roughness with root-mean square (RMS) values of typically less than 10 nm can neither be characterized adequately nor can they be resolved by realistic FDTD grids. In contrast to that, incompletely etched coupling gaps and etch undercut are systematic deviations. If they are known to a sufficient degree of accuracy, they can in principle be taken into account by the simulation. We have attempted to estimate the geometrical parameters at certain points of the real structure from SEM pictures, and we have adapted the simulated structure to these findings. Since the accuracy of SEM-based geometrical characterization is limited, we change the parameters in the simulation only by multiples of the corresponding Yee cell size. When increasing the width of the coupling gaps by two Yee cells ($g_1 = 0.200\,\mu$m and $g_2 = 0.250$ μm) and simultaneously reducing the waveguide width by one Yee cell ($w = 0.375\,\mu$m), the entire resonance comb shifts up by 1.5 THz, and there is no compliance with the measurement any more. When only a widening of the coupling gaps is considered ($g_1 = 0.200\,\mu$m and $g_2 = 0.250$ μm), the change in resonance frequencies compared to the original structure is negligible, but the FWM conversion efficiency is increased. Coupling gaps that are wider than the nominal values apparently yield a structure that is closer to critical coupling conditions. However, this effect will be counteracted by the influence of surface roughness and of incompletely etched coupling gaps: Using gaps of the aforementioned widths but with a depth of only 1 μm reduces both the Q-factor of the resonator and the FWM efficiency dramatically. Since the effective refractive index of the fundamental waveguide mode $n_e \approx 3.04$ is less than the refractive index of the InP substrate, residual material within the gaps not only leads to stronger coupling between the ring and the bus waveguide, but also makes the resonator lossy by allowing light to couple into the substrate.

All attempts to adapt the parameters of the simulated structure thus lead to changes which are qualitatively plausible, but quantitatively too extreme to comply with the measurement. This is due to the fact that SEM-based characterization only allows the investigation of a few points within the structure. Our knowledge of the real device's fully three-dimensional geometry is therefore always incomplete. Obviously, even subtle differences between the real-world and the simulated structure can lead to significant deviations in the device's behaviour. The deviations between the measurement and the benchmark simulation using nominal geometrical parameters are thus well within the expected range, and we may state that the predictive power of the benchmark model is mainly limited by an incomplete knowledge of the real device's geometry.

9.5.2   Simplified FDTD-Models

As concluded in the last section, the benchmark simulation gives accurate results if the geometry of the simulated structure is exactly known. It therefore makes sense to evaluate the simplified FDTD models by comparing their predictions to the results obtained from this reference simulation. For the linear model it can be seen from Fig. 9.6 that a 3D simulation that incorporates material dispersion is necessary to predict the correct FSR to within 1 % of accuracy. We further find that using a 2D model with material dispersion gives better results than a 3D model without material dispersion. As a rule of thumb, we may state that this holds for all waveguide systems in which material dispersion dominates over waveguide dispersion. This is a particularly interesting result, since it shows that in this case a compact but advanced 2D simulation performs better than a less developed computationally expensive 3D simulation.

From Fig. 9.6 it is also obvious that the EDC technique has only a marginal effect on the positions of the resonances. It is nevertheless an important extension of the FDTD model: Fig. 9.5 shows the 2D FDTD spectrum without EDC (2D FDTD). It can be seen that the peak value near 192 THz is suppressed by approximately 3 dB compared to the other peaks. Similar artefacts have been observed also in spectra for other simulations in which the staircasing error has not been corrected. The suppression is due to radiation and reflection loss caused by the staircased model of the curved waveguide contour.

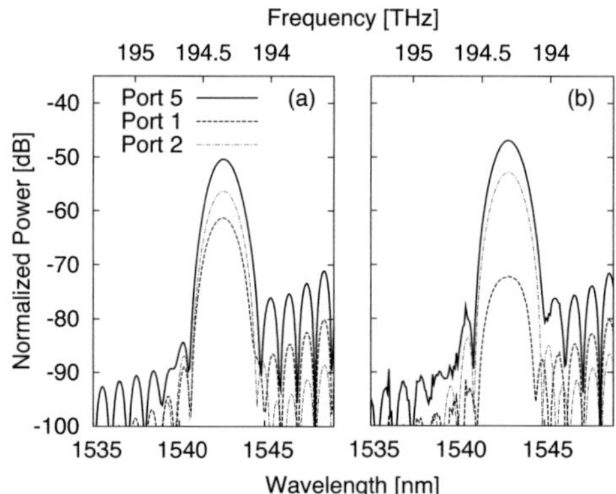

Fig. 9.9. Comparison of the converted power levels at different ports for a maximum pump electric field $E_p = 5.65 \times 10^6$ V/m ($P_p \approx 13$ dBm) (a) 3D FDTD including material dispersion, but no EDC (b) 3D FDTD including material dispersion and EDC.

Another error caused by staircasing is evident from Fig. 9.2, which shows the FWM spectra at different ports for a simulation without EDC: A converted wave at approximately 1 540 nm can be detected at the input port 1, even though no field was launched at this frequency. This is caused by back-reflections from inside the ring. Fig. 9.9 (a) shows again the converted lines at the different ports for a simulation without EDC smoothing, and Fig. 9.9 (b) shows the same lines obtained from a simulation with EDC. The converted spectral component at the input port 1 decreases by approximately 11 dB once EDC smoothing is applied. We may therefore state that for the case where EDC smoothing was omitted, the back-reflections from inside the ring are indeed caused predominantly by staircasing. For the strongly guiding high index-contrast waveguides used here, reflections due to mode mismatch at the transitions between the straight and the bent sections of the ring are thus weaker by an order of magnitude.

The reflection and scattering due to "numerical" staircasing roughness can be considered as an additional source of resonator loss for the case where the EDC technique is not applied. This lowers the field enhancement and thus the conversion efficiency as can be seen in Fig. 9.10, where we plot the conversion efficiency against the pump power for the different FDTD models: The results clearly separate into two bands — one for the simulations with EDC correction (high conversion efficiency), and one for the simulations without (lower conversion efficiency). The agreement of the measurement and the simulations without EDC must be considered coincidental, since the staircasing errors are accidentally equal to the deviations caused by the fabrication tolerances of the device.

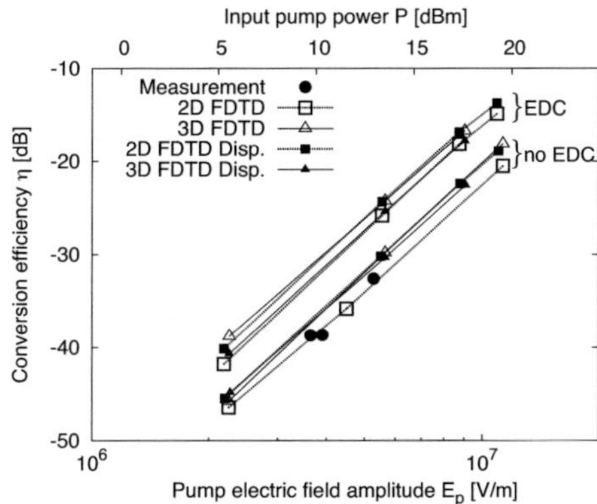

Fig. 9.10. Comparison of the conversion efficiencies obtained by the measurement and the FDTD analyses with and without material dispersion and/or EDC-smoothing.

We conclude that the EDC smoothing is essential for the correct prediction of the field enhancement, the conversion efficiency and all other parameters that depend on the resonator loss, e. g., the Q-factor, the finesse, and the linewidth of the resonance. Interestingly, the results for the 2D and the 3D FDTD agree relatively well. This implies that, again, an advanced 2D model with EDC is more powerful than a computationally expensive 3D model without EDC.

9.6 Summary

We have demonstrated a fully 3D FDTD model of FWM in a racetrack resonator, incorporating features like material dispersion and EDC correction of staircasing errors. We have investigated its accuracy by comparison with experimental results. Using an analytical formulation which allows us to separate different physical effects in a systematic way, discrepancies between the simulation and the experiment were attributed to an inaccurate representation of the real structure.

By investigating simplified versions of the benchmark model, we were able to evaluate the necessity of the various features. Specifically, we have found that a 3D model incorporating material dispersion is necessary to accurately model the dispersion relation of the high index-contrast waveguide. With respect to resonant field enhancement, it turns out that staircasing artefacts cause a large error in predictions of quantities that depend on the resonator losses. If these parameters are of interest, the EDC technique (or a similar technique which eliminates staircasing errors) must be used. Generally, incorporating features like EDC smoothing and material dispersion into a compact 2D model does not require excessive computational resources and gives much better results than using a computationally expensive 3D model without these features.

Chapter 10

Summary and Future Work

In this thesis, nanophotonic waveguide devices have been investigated for optical signal processing applications. The high index-contrast nature of these devices requires a reformulation of linear and nonlinear mode coupling relations that are known from low index-contrast (fiber) waveguides, see Chapter 1. The theoretical investigations are supplemented by experiments – see Chapter 2 for a detailed description of the setups. In the following, we shortly summarize the achievements and develop suggestions for future research.

Waveguide Roughness: Based on a novel semianalytical calculation of the radiation modes in three dimensions, we investigate roughness-induced coupling of power from guided to radiation modes, see Chapter 3. The results are confirmed by numerical simulations and measurements, and design rules for low-loss straight waveguides are derived.

As a next step, the semianalytical formulation should be used to develop a generalized measure of sidewall roughness taking into account both the amplitude and the correlation length of the sidewall perturbations. By characterizing different waveguide fabrication processes, their ability to produce low-loss high index-contrast (HIC) devices can be rated. Roughness losses can then be predicted and considered during the design process. This will considerably reduce trial-and-error approaches in the design of HIC waveguide devices.

Waveguide Bends: Based on the understanding of roughness-induced scattering loss, the concept of optimized device contours is introduced in Chapter 4. This novel design approach is applied to waveguide bends.

As a next step, a family of basic passive waveguide devices with ideal contours, e.g., multi-mode interference couplers (MMI), should be designed and tested. A library of ideal standard devices could considerably speed up the design of complex low-loss planar lightwave circuits (PLC).

Preferential Etching of Waveguides: A novel, highly precise fabrication method for special waveguide geometries is introduced in Chapter 5. It is based on preferential wet etching of crystalline materials. We have fabricated promising prototype structures.

As a next step, such structures should be combined with waveguide devices that are obtained by conventional anisotropic dry etching. Working with VLSI-grade chemicals under strict cleanroom conditions should then yield record low-loss waveguides.

Electro-Optic Modulators: In Chapter 6, novel silicon-based modulator schemes with unprecedented electrical bandwidth and compactness are proposed and investigated. The fabrication and characterization of prototype structures is envisaged for the near future. Incorporating the device into a silicon-based transceiver could be subject of joint research projects together with industrial partners.

Nonlinear Waveguides: Third-order nonlinear silicon-based strip and slot waveguides for all-optical signal processing are discussed Chapter 7. The fabrication of prototype structures is in progress. The next step should be an experimental proof-of-principle of ultrafast all-optical wavelength conversion.

Quantum Dot Semiconductor Optical Amplifiers: Nonlinear dynamics of an InAs/GaAs quantum dot (QD) amplifier are investigated, Chapter 8, and low alpha-factors have been measured. Further investigations on the influence of the device length, the temperature, the pump current, and the pump impulse energy are currently being performed. As a next step, an impulse-shaper shall be incorporated into the pump arm of the setup to investigate the dynamics for different shapes of the pump impulse. It would further be of high interest to measure the spectral width of the nonlinear gain change and phase change by performing phase-sensitive time-resolved spectroscopy. From the theoretical side, a more accurate model of the device should allow us to quantify the effects of two-photon absorption (TPA), which can then be exploited as an additional mechanism for refilling the depleted states of the QD. This is expected to mitigate pattern effects.

Nonlinear Resonators: Resonant enhancement as discussed in Chapter 9 can significantly reduce power levels for nonlinear interaction. Incorporating nonlinear strip and slot waveguides (Chapter 7) into resonant devices may therefore be key towards chip-scale ultra-fast optical signal processing.

This work was supported by the Center for Functional Nanostructures (CFN) of the Deutsche Forschungsgemeinschaft (DFG) within projects A3.1 and A4.4, and by the European project TRIUMPH (grant IST-027638 STP).

Appendix A

Mathematical Definitions and Conventions

To clarify notation, we summarize fundamental relations and definitions that are used in this thesis.

A.1 Time- and Frequency-Domain Quantities

If not otherwise stated, t is the independent time domain variable, and ω denotes the corresponding frequency domain variable. The quantities $u(t)$, $v(t)$, and $h(t)$ are functions in the time domain, and $\widetilde{u}(\omega)$, $\widetilde{v}(\omega)$, and $\widetilde{h}(\omega)$ are the corresponding frequency domain spectra.

A.1.1 Fourier Transformation

The Fourier transform of a function $u(t)$ with respect to the independent variable t is denoted as $\mathfrak{F}_t\left\{u(t)\right\}$. Accordingly, $\mathfrak{F}_\omega^{-1}\left\{\widetilde{u}(\omega)\right\}$ refers to the inverse Fourier transform of a function $\widetilde{u}(\omega)$ with respect to the independent variable ω,

$$\mathfrak{F}_t\left\{u(t)\right\} = \widetilde{u}(\omega) = \int\limits_{-\infty}^{+\infty} u(t)\,\mathrm{e}^{-\mathrm{j}\omega t}\,\mathrm{d}t, \tag{A.1}$$

$$\mathfrak{F}_\omega^{-1}\left\{\widetilde{u}(\omega)\right\} = u(t) = \frac{1}{2\pi}\int\limits_{-\infty}^{+\infty} \widetilde{u}(\omega)\,\mathrm{e}^{\mathrm{j}\omega t}\,\mathrm{d}\omega. \tag{A.2}$$

The independent variable t is usually represents the time, and ω is angular frequency.

A.1.2 Convolution

The convolution of two time domain functions $u(t)$ and $v(t)$ is given by

$$u\left(t\right) * v\left(t\right) = \int_{-\infty}^{+\infty} u(\tau)v(t-\tau)\,\mathrm{d}\tau. \tag{A.3}$$

In the frequency domain, this corresponds to a simple multiplication,

$$\mathfrak{F}_t\left\{u\left(t\right) * v\left(t\right)\right\} = \mathfrak{F}_t\left\{u\left(t\right)\right\}\mathfrak{F}_t\left\{v\left(t\right)\right\}. \tag{A.4}$$

A.1.3   Analytic Signals and Hilbert Transform

The imaginary part of an analytic signal $\underline{x}(t)$ is the Hilbert transform of its real part,

$$\underline{u}(t) = u(t) + \mathrm{j}\,\mathfrak{H}_t\left\{u(t)\right\}, \tag{A.5}$$

where the Hilbert transform with respect to time is defined as

$$\mathfrak{H}_t\left\{u(t)\right\} = \frac{1}{\pi}\mathfrak{P}\int_{-\infty}^{+\infty}\frac{u(\tau)}{t-\tau}\,\mathrm{d}\tau. \tag{A.6}$$

$\mathfrak{P}$ denotes the Cauchy principal value of the integral. The Hilbert-transformation can be interpreted as a convolution in the time-domain,

$$\mathfrak{H}_t\left\{u(t)\right\} = \frac{1}{\pi}\mathfrak{P}\left\{\frac{1}{t}\right\} * u(t), \tag{A.7}$$

which corresponds to a multiplication in the Fourier-domain

$$\mathfrak{F}_t\left\{\mathfrak{H}_t\left\{u(t)\right\}\right\} = \mathfrak{F}_t\left\{\frac{1}{\pi}\mathfrak{P}\left\{\frac{1}{t}\right\}\right\}\mathfrak{F}_t\left\{u(t)\right\}.$$

Using the Fourier-transform of $\frac{1}{\pi}\mathfrak{P}\left\{\frac{1}{t}\right\}$, we find

$$\mathfrak{F}_t\left\{\frac{1}{\pi}\mathfrak{P}\left\{\frac{1}{t}\right\}\right\} = \mathrm{j}\left(2H\left(-\omega\right)-1\right) = -\mathrm{j}\operatorname{sgn}\left(\omega\right), \tag{A.8}$$

where $H\left(\cdot\right)$ is the Heaviside function[1] and $\operatorname{sgn}\left(\cdot\right)$ is the sign function.

The Fourier transform $\underline{\widetilde{u}}(\omega)$ of the analytical signal $\underline{u}(t)$ can be written as

$$\underline{\widetilde{u}}(\omega) = 2H\left(\omega\right)\widetilde{u}(\omega). \tag{A.9}$$

As a special case of practical interest, we have to consider a signal $u(t)$ which has a continuous wave (cw) component, i.e. $u(t) = C$ and $\widetilde{u}(\omega) = C\delta(\omega)$, where $\delta(\omega)$ denotes

---

[1] The Heaviside $H(x)$ function is a generalized function defined by a distribution $\int_{-\infty}^{\infty} H(x)\Phi(x)\,\mathrm{d}x = \int_{0}^{\infty}\Phi(x)\,\mathrm{d}x$ for any "good" test function $\Phi(x)$. For the formal definition of a "good" function, see [51, Definition 2.1].

the Dirac delta function[2]. The single-sided spectrum can then be written formally as $\widetilde{\underline{u}}(\omega) = 2C\,H(\omega)\,\delta(\omega)$. We note that products of generalized functions like the Dirac delta function $\delta(\omega)$ and the Heaviside function $H(\omega)$ are only defined if both functions fulfill certain conditions [51, Chapter 6]. Since $\delta(\omega)$ is an even function of ω, the definition

$$\int_{-\infty}^{+\infty} H(\omega)\,\delta(\omega)\,\Phi(\omega)\,\mathrm{d}\omega = \int_0^\infty \delta(\omega)\,\Phi(\omega)\,\mathrm{d}\omega = \frac{1}{2}\Phi(0) \tag{A.10}$$

is commonly used. As a consequence, the delta-impulse at $\omega = 0$ is not changed when going from the double-sided to the single-sided spectrum,

$$\widetilde{u}(\omega) = C\,\delta(\omega) \implies \widetilde{\underline{u}}(\omega) = C\,\delta(\omega). \tag{A.11}$$

Table (A.1.3) summarizes the relations between the different time domain and frequency domain quantities.

A.1.4 Correlations

When defining correlations of deterministic time domain functions $u(t)$ and $v(t)$, we have to distinguish whether or not the functions are L^2-integrable, i.e. whether or not their energy is finite. If so, we use the definition for energy signals,

$$R_{uv}(\tau) = \left(u(t) * v^*(-t)\right)(\tau) = \int_{-\infty}^{+\infty} u(t)\,v^*(t-\tau)\,\mathrm{d}t. \tag{A.12}$$

For signals of infinite energy, e.g. periodic signals, the convention

$$R_{uv}(\tau) = \lim_{T\to\infty} \frac{1}{2T} \int_{-T}^{T} u(t)\,v^*(t-\tau)\,\mathrm{d}t \tag{A.13}$$

is used. In the case of weakly stationary stochastic processes $u(t)$ and $v(t)$, the correlation function is based on an ensemble average denoted by $\langle\cdot\rangle$,

$$R_{uv}(\tau) = \langle u(t)\,v^*(t-\tau)\rangle. \tag{A.14}$$

For processes that are ergodic (and thus strictly stationary), the ensemble average in Eq. (A.14) can be replaced by a integral over time similar to Eq.(A.13).

---

[2]The Dirac delta function $\delta(x)$ is a generalized function defined by a distribution $\int_{-\infty}^{\infty} \delta(x)\Phi(x)\,\mathrm{d}x = \Phi(0)$ for any "good" test function $\Phi(x)$. For the formal definition of a "good" function, see [51, Definition 2.1].

	$u(t)$	$\underline{u}(t)$	$\widetilde{u}(\omega)$	$\underline{\widetilde{u}}(\omega)$
$u(t)$		$= \mathrm{Re}\,\{\underline{u}(t)\}$	$= \frac{1}{2\pi} \int\limits_{-\infty}^{+\infty} \widetilde{u}(\omega)\,\mathrm{e}^{\mathrm{j}\omega t}\,\mathrm{d}\omega$	$= \mathrm{Re}\left\{ \frac{1}{2\pi} \int\limits_{-\infty}^{+\infty} \underline{\widetilde{u}}(\omega)\,\mathrm{e}^{\mathrm{j}\omega t}\,\mathrm{d}\omega \right\}$
$\underline{u}(t)$	$= u(t) - \mathrm{j}\,\mathfrak{H}_t\,\{u(t)\}$		$= \frac{1}{\pi} \int\limits_{0}^{\infty} \underline{\widetilde{u}}(\omega)\,\mathrm{e}^{\mathrm{j}\omega t}\,\mathrm{d}\omega$	$= \frac{1}{2\pi} \int\limits_{-\infty}^{+\infty} \underline{\widetilde{u}}(\omega)\,\mathrm{e}^{\mathrm{j}\omega t}\,\mathrm{d}\omega$
$\widetilde{u}(\omega)$	$= \int\limits_{-\infty}^{+\infty} u(t)\,\mathrm{e}^{-\mathrm{j}\omega t}\,\mathrm{d}\omega$	$= \int\limits_{-\infty}^{+\infty} \mathrm{Re}\,\{\underline{u}(t)\}\,\mathrm{e}^{-\mathrm{j}\omega t}\,\mathrm{d}\omega$		$= H\,(\omega)\,\underline{\widetilde{u}}(\omega)$ $+ H\,(-\omega)\,\underline{\widetilde{u}}^{*}(-\omega)$
$\underline{\widetilde{u}}(\omega)$	$= 2H(\omega) \int\limits_{-\infty}^{+\infty} u(t)\,\mathrm{e}^{-\mathrm{j}\omega t}\,\mathrm{d}t$	$= \int\limits_{-\infty}^{+\infty} \underline{u}(t)\,\mathrm{e}^{-\mathrm{j}\omega t}\,\mathrm{d}t$	$= 2H\,(\omega)\,\widetilde{u}(\omega)$	

Table A.1. Relationships between real and analytic time-domain signals and their spectra

A.1.5   Spectral Densities

For L^2-integrable functions, the spectral energy density is obtained as the Fourier transform of the autocorrelation function,

$$S_{uv}(\omega) = \mathfrak{F}_t\left\{R_{uv}(\tau)\right\}(\omega) = \widetilde{u}(\omega)\,\widetilde{v}^*(\omega). \tag{A.15}$$

Similarly, for functions that are not L^2-integrable, i.e. for signals of infinite energy, the Fourier transform of the correlation function yields the spectral power density. For weakly stationary stochastic processes, this is made explicit by the Wiener-Khintchine theorem [83],

$$S_{uv}(\omega) = \mathfrak{F}_t\left\{R_{uv}(\tau)\right\}(\omega). \tag{A.16}$$

In contrast to Eq. (A.15), the spectral power density in Eq. (A.16) cannot be represented by the product of the Fourier transforms of the respective functions[3]. We note that for weakly stationary stochastic processes, Eq. (A.16) is only valid if the integral $\int_{-\infty}^{+\infty}|R_{uv}(\tau)\tau|\,\mathrm{d}\tau$ is bounded [27].

A.2   Linear Time-Variant Systems

The output signal $v(t)$ of a system is related to its input signal $u(t)$ by the system operator S. For linear time-invariant (LTI) systems, S does not depend on time. Input $u(t)$ and output $v(t) = S\{u(t)\}$ are related by a simple convolution with the system's impulse response $h(t)$ in the time-domain. This corresponds to a multiplication with the transfer function $\widetilde{h}(\omega) = \mathfrak{F}_t\{h(t)\}$ in the frequency domain. In contrast to that, linear time-variant (LTV) systems have a time-dependent operator S_t and can be described by an explicitly time-dependent Green's function $g(t-\tau,\tau)$,

$$v(t) = S_t\{u(t)\} = \int\limits_{-\infty}^{+\infty} u(\tau)\,g(t-\tau,\tau)\,\mathrm{d}\tau. \tag{A.17}$$

The time-domain Green's function $g(t-\tau,\tau) = S_t\{\delta(t-\tau)\}$ represents the response of the system to a Dirac delta excitation $\delta(t-\tau)$ at $t=\tau$. The argument τ defines the position of the excitation impulse on the time axis, and $t-\tau$ is the time that has passed after the excitation. For causal systems, $g(t-\tau,\tau) = 0$ for $t < \tau$. In the frequency domain, the output spectrum $\widetilde{v}(\omega)$ is related to the input spectrum $\widetilde{u}(\omega)$ by

$$\widetilde{v}(\omega) = \frac{1}{2\pi}\int\limits_{-\infty}^{+\infty} \widetilde{u}(\xi)\,\widetilde{g}(\omega,\omega-\xi)\,\mathrm{d}\xi. \tag{A.18}$$

The Green's function $\widetilde{g}(\omega,\omega-u)$ in the frequency domain is obtained from $g(t-\tau,\tau)$ by means of Fourier transformation,

$$\widetilde{g}(\omega_1,\omega_2) = \iint\limits_{-\infty}^{+\infty} g(t_1,t_2)\,\mathrm{e}^{-\mathrm{j}(\omega_1 t_1 + \omega_2 t_2)}\,\mathrm{d}t_1\,\mathrm{d}t_2. \tag{A.19}$$

---

[3]For signals of infinite energy, the Fourier transform can be a generalized function – it can e.g. contain δ-functions. The product of generalized functions is however undefined.

Appendix B

Linear and Nonlinear Polarization

B.1 The Optical Susceptibility Tensor

In Eq. (1.7), we have introduced the time-domain relation between the polarization $\mathbf{P}(\mathbf{r}, t)$ and the electric field $\mathbf{E}(\mathbf{r}, t)$,

$$\mathbf{P}(\mathbf{r}, t) = \sum_n \mathbf{P}^{(n)}(\mathbf{r}, t),$$

where the n-th order polarization $\mathbf{P}^{(n)}(\mathbf{r}, t)$ is defined by

$$\mathbf{P}^{(n)}(\mathbf{r}, t) = \epsilon_0 \int_{-\infty}^{+\infty} \cdots \int \underline{\chi}^{(n)} \left(t - \tau_1, t - \tau_2, \ldots, t - \tau_n \right) \tag{B.1}$$

$$\vdots \ \mathbf{E}(\mathbf{r}, \tau_1) \mathbf{E}(\mathbf{r}, \tau_2) \ldots \mathbf{E}(\mathbf{r}, \tau_2) \ \mathrm{d}\tau_1 \ldots \mathrm{d}\tau_n.$$

The time-domain Volterra kernel of n-th order, $\underline{\chi}^{(n)}(t_1, t_2, \ldots, t_n)$, is a tensor of rank $n+1$. Omitting the spatial dependence of all quantities for the sake of readability, Eq. (B.1) can be rewritten for the different vector components $q_0, q_1 \ldots, q_n \in \{x, y, z\}$,

$$P_{q_0}^{(n)}(t) = \epsilon_0 \sum_{\substack{q_1, q_2 \ldots, q_n \\ \in \{x,y,z\}}} \int_{-\infty}^{+\infty} \cdots \int \chi_{q_0:q_1 q_2 \ldots q_n}^{(n)} \left(t - \tau_1, t - \tau_2, \ldots, t - \tau_n \right)$$

$$\times E_{q_1}(\tau_1) E_{q_2}(\tau_2) \ldots E_{q_n}(\tau_n) \ \mathrm{d}\tau_1 \ldots \mathrm{d}\tau_n. \tag{B.2}$$

The frequency-domain analogue of Eq. (B.2) reads

$$\widetilde{P}_{q_0}^{(n)}(\omega) = \frac{\epsilon_0}{(2\pi)^{n-1}} \sum_{\substack{q_1, q_2 \ldots, q_n \\ \in \{x,y,z\}}} \int_{-\infty}^{+\infty} \cdots \int \widetilde{\chi}_{q_0:q_1 q_2 \ldots q_n}^{(n)} \left(\omega : \omega_1, \omega_2, \ldots, \omega_{n-1}, \omega - \sum_{m=1}^{n-1} \omega_m \right)$$

$$\times \widetilde{E}_{q_1}(\omega_1) \widetilde{E}_{q_2}(\omega_2) \ldots \widetilde{E}_{q_{n-1}}(\omega_{n-1}) \widetilde{E}_{q_n}\left(\omega - \textstyle\sum_{m=1}^{n-1}\omega_m\right) \mathrm{d}\omega_1 \ldots \mathrm{d}\omega_{n-1}, \tag{B.3}$$

where the components $\widetilde{\chi}^{(n)}_{q_0:q_1 q_2 \ldots q_n}$ of the n-th order susceptibility tensor $\underline{\widetilde{\chi}}^{(n)}$ are defined by

$$\widetilde{\underline{\chi}}^{(n)}_{q_0:q_1 q_2 \ldots q_n} \left(\sum_{m=1}^{n} \omega_m : \omega_1, \omega_2, \ldots, \omega_n \right) \tag{B.4}$$

$$= \int_{-\infty}^{+\infty} \cdots \int \chi^{(n)}_{q_0:q_1 q_2 \ldots q_n} \left(\tau_1, \tau_2, \ldots, \tau_n \right) e^{-j\omega_1 \tau_1} e^{-j\omega_2 \tau_2} \ldots e^{-j\omega_n \tau_n} \, d\tau_1 \, d\tau_2 \ldots d\tau_n.$$

All integrals over ω or τ extend from $-\infty$ to $+\infty$.

The short-form notation for the tensor product used in Eq. (B.1) already implies the summation over the vector components,

$$\underline{\chi}^{(n)} \left(t - \tau_1, t - \tau_2, \ldots, t - \tau_n \right) \, \vdots \, \mathbf{E}\left(\tau_1\right) \mathbf{E}\left(\tau_2\right) \ldots \mathbf{E}\left(\tau_n\right) \tag{B.5}$$

$$= \sum_{\substack{q_0, q_1 \ldots, q_n \\ \in \{x,y,z\}}} e_{q_0} \chi^{(n)}_{q_0:q_1 q_2 \ldots q_n} \left(t - \tau_1, t - \tau_2, \ldots, t - \tau_n \right) E_{q_1}\left(\tau_1\right) E_{q_2}\left(\tau_2\right) \ldots E_{q_n}\left(\tau_n\right).$$

Transferring this definition to the frequency domain, we obtain

$$\underline{\widetilde{\chi}}^{(n)} \left(\omega_\Sigma : \omega_1, \omega_2, \ldots, \omega_{n-1}, \omega_n \right) \, \vdots \, \widetilde{\mathbf{E}}\left(\omega_1\right) \widetilde{\mathbf{E}}\left(\omega_2\right) \ldots \widetilde{\mathbf{E}}\left(\omega_{n-1}\right) \widetilde{\mathbf{E}}\left(\omega_n\right) \tag{B.6}$$

$$= \sum_{\substack{q_0, q_1 \ldots, q_n \\ \in \{x,y,z\}}} e_{q_0} \widetilde{\chi}^{(n)}_{q_0:q_1 q_2 \ldots q_n} \left(\omega : \omega_1, \omega_2, \ldots, \omega_{n-1}, \omega_n \right) \widetilde{E}_{q_1}\left(\omega_1\right) \widetilde{E}_{q_2}\left(\omega_2\right) \ldots E_{q_n}\left(\omega_n\right),$$

where $\omega_\Sigma = \sum_{m=1}^{n} \omega_m$ denotes the sum frequency. Using this definition, the frequency-domain analogue of Eq. (B.1) reads

$$\widetilde{\mathbf{P}}^{(n)}\left(\omega\right) = \frac{\epsilon_0}{\left(2\pi\right)^{n-1}} \int_{-\infty}^{+\infty} \cdots \int \underline{\widetilde{\chi}}^{(n)} \left(\omega : \omega_1, \omega_2, \ldots, \omega_{n-1}, \omega - \sum_{m=1}^{n-1} \omega_m \right)$$

$$\vdots \, \widetilde{\mathbf{E}}\left(\omega_1\right) \widetilde{\mathbf{E}}\left(\omega_2\right) \ldots \widetilde{\mathbf{E}}\left(\omega_{n-1}\right) \widetilde{\mathbf{E}}\left(\omega - \sum_{m=1}^{n-1} \omega_m\right) d\omega_1 \ldots d\omega_{n-1}. \tag{B.7}$$

B.2 Properties of the Susceptibility Tensor

B.2.1 Causality

A detailed discussion on the properties of the susceptibility tensor can be found in standard textbooks on nonlinear optics, e.g. [14]. Here, we only give a short summary.

In linear optics, the causality of the dielectric impulse response can be directly translated to the frequency domain and gives a useful relation between the real and the imaginary part of the complex optical susceptibility. The time-domain formulation of the nonlinear optical susceptibility must also obey the causality principle, i.e.

$$\chi^{(n)}_{q_0:q_1 q_2 \ldots q_n} \left(\tau_1, \tau_2, \ldots, \tau_n \right) = 0 \quad \text{for} \quad \tau_1 < 0 \, \vee \, \tau_2 < 0 \cdots \vee \, \tau_n < 0. \tag{B.8}$$

However, the frequency-domain formulation of the causality condition is somewhat intricate. For some nonlinear processes, Kramers-Kronig relations similar to the linear case are valid, but for some processes (e.g. SPM), it is not possible to formulate a Kramers-Kronig relation [14].

B.2.2 Symmetries

The different components of the nonlinear optical susceptibility are not independent from each other, and even without further assumptions, they possess certain symmetry properties [14], as listed below. In the following $\omega_\Sigma = \sum_{m=1}^{n} \omega_m$ denotes the sum of the input frequencies.

- Time-domain field quantities have to be real, This relates positive- and negative-frequency components:

$$\underline{\widetilde{\chi}}^{(n)}_{q_0:q_1 q_2 \dots q_n} (\omega_\Sigma : \omega_1, \omega_2, \dots, \omega_n) = \left[\underline{\widetilde{\chi}}^{(n)}_{q_0:q_1 q_2 \dots q_n} (-\omega_\Sigma : -\omega_1, -\omega_2, \dots, -\omega_n) \right]^* \quad (\text{B.9})$$

- To obtain a physically relevant time-domain quantity, we have to perform both an integral over all frequency arguments $\omega_1 \dots \omega_n$ from $-\infty$ to $+\infty$ and a sum where each of the indices $q_1 \dots q_n$ assumes the values x, y and z. The corresponding expression in Eq. (B.4) is not changed if we swap two of the frequency arguments $\omega_1, \dots, \omega_n$ and the corresponding vector component indices simultaneously. The corresponding tensor elements can thus be assumed to be identical:

$$\underline{\widetilde{\chi}}^{(n)}_{q_0:q_1 \dots q_i q_j \dots q_n} (\omega_\Sigma : \omega_1, \dots, \omega_i, \omega_j, \dots, \omega_n) = \underline{\widetilde{\chi}}^{(n)}_{q_0:q_1 \dots q_j q_i \dots q_n} (\omega_\Sigma : \omega_1, \dots, \omega_j, \omega_i, \dots, \omega_n)$$
$$(\text{B.10})$$

The intrinsic permutation symmetry holds for all pairs of frequencies and the corresponding indices except for the resulting frequency ω_Σ and the resulting vector component index q_0.

- For lossless media, two more symmetry properties apply:

1. All of the components $\underline{\widetilde{\chi}}^{(n)}_{q_0:q_1 q_2 \dots q_n}$ are real,

$$\underline{\widetilde{\chi}}^{(n)}_{q_0:q_1 q_2 \dots q_n} (\omega_\Sigma : \omega_1, \omega_2, \dots, \omega_n) = \left[\underline{\widetilde{\chi}}^{(n)}_{q_0:q_1 q_2 \dots q_n} (\omega_\Sigma : \omega_1, \omega_2, \dots, \omega_n) \right]^*. \quad (\text{B.11})$$

2. The permutation symmetry according to Eq. (B.10) also holds for the resulting frequency ω. Since this frequency is always the sum of all other frequency arguments, signs must be changed appropriately when interchanging the first argument with any other argument,

$$\underline{\widetilde{\chi}}^{(n)}_{q_0:q_1 \dots q_i \dots q_n} (\omega_\Sigma : \omega_1, \dots, \omega_i, \dots, \omega_n) = \underline{\widetilde{\chi}}^{(n)}_{q_i:q_1 \dots q_0 \dots q_n} (\omega_i : -\omega_1, \dots, \omega_\Sigma, \dots, -\omega_n)$$
$$(\text{B.12})$$

- In many cases of practical interest, optical media are operated at frequencies far below their lowest resonance frequency. In this case, the medium can not only assumed to be lossless, but the nonlinear susceptibility is essentially frequency independent within the considered wavelength range. The frequency arguments can then be permuted without permuting the indices,

$$\underline{\widetilde{\chi}}^{(n)}_{q_0:q_1\ldots q_i q_j\ldots q_n}\left(\omega_\Sigma:\omega_1,\ldots,\omega_i,\omega_j,\ldots,\omega_n\right) = \underline{\widetilde{\chi}}^{(n)}_{q_0:q_1\ldots q_i q_j\ldots q_n}\left(\omega_\Sigma:\omega_1,\ldots,\omega_j,\omega_i,\ldots,\omega_n\right),$$

$$(B.13)$$

for any two indices q_2 and q_1. This symmetry property is referred to as Kleinman's Symmetry.

- The number of independent components of the susceptibility tensor can be further reduced if the material system possesses spatial symmetries. For the second- and the third-order nonlinear susceptibility, the nonvanishing tensor elements and their relations are listed for each of the 32 crystal classes in [14].

B.3   Linear and Nonlinear Polarization in SVEA

Most interesting for technical applications are the second- and third-order nonlinear susceptibilities and their applications in optical signal processing. Since optical carriers vary usually much faster than any (electric) modulation signal, the slowly-varying envelope approximation (SVEA) is an excellent method of analyzing optical signals. We will derive specialized equations for second- and third-order processes that describe the nonlinear optical interaction in terms of signal envelopes. The total electric field consists of different carriers at frequencies ω_ν (optical channels), with slowly-varying envelopes $\widehat{\underline{\mathbf{E}}}(\mathbf{r}, t, \omega_\nu)$,

$$\mathbf{E}(\mathbf{r}, t) = \frac{1}{2}\sum_{\nu=-N}^{N}(1+\delta_{\nu,0})\,\widehat{\underline{\mathbf{E}}}(\mathbf{r}, t, \omega_\nu)\,\mathrm{e}^{\mathrm{j}\omega_\nu t}, \tag{B.14}$$

where $\omega_{-\nu} = -\omega_\nu$ for $\nu \neq 0$ and $\omega_0 = 0$. Analogous relations hold for the magnetic field. Quantities with negative or zero frequency arguments can be obtained from their positive-frequency counterparts, and are given by

$$\widehat{\underline{\mathbf{E}}}(\mathbf{r}, t, \omega_{-\nu}) = \widehat{\underline{\mathbf{E}}}^{*}(\mathbf{r}, t, \omega_\nu), \qquad \widehat{\underline{\mathbf{E}}}(\mathbf{r}, t, \omega_0) \in \mathbb{R}^3. \tag{B.15}$$

A special form of the SVEA is based on the expansion in waveguide modes. For the sake of simplicity, we consider only the guided modes. Nonlinear interaction with and among radiation modes can safely be neglected in a nonlinear analysis, since the corresponding power densities are usually orders of magnitude too small to show any nonlinear effect. Each frequency component excites different modes indicated by a summation over the mode index μ. The signal envelope centered at ω_ν and propagating in the μ-th mode is labelled $A_\mu(z, t, \omega_\nu)$. The SVEA is thus given by

$$\widehat{\underline{\mathbf{E}}}(\mathbf{r}, t, \omega_\nu) = \sum_{\mu}(1+\delta_{\nu,0})\,A_\mu(z, t, \omega_\nu)\frac{\mathcal{E}_\mu(x, y, \omega_\nu)}{\sqrt{\mathcal{P}_\mu(\omega_\nu)}}\,\mathrm{e}^{-\mathrm{j}\beta_\mu(\omega_\nu)z} \tag{B.16}$$

An analogous relation holds for the magnetic field. Quantities with negative or zero frequency arguments are given by

$$
\begin{aligned}
\beta_\mu\left(-\omega_\nu\right) &= \beta_\mu^*\left(\omega_\nu\right), & \beta_\mu\left(0\right) &= 0, \\
\mathcal{E}_\mu(x,y,\omega_{-\nu}) &= \mathcal{E}_\mu^*(x,y,\omega_\nu), & \mathcal{E}_\mu(x,y,0) &\in \mathbb{R}, \\
\widetilde{A}_\mu(z,\omega,\omega_{-\nu}) &= \widetilde{A}_\mu^*(z,-\omega,\omega_\nu), & \widetilde{A}_\mu(\mathbf{r},\omega,0) &= \widetilde{A}_\mu^*(\mathbf{r},-\omega,0).
\end{aligned}
\tag{B.17}
$$

B.3.1 Linear Polarization

Specializing Eq. (B.7) for the linear polarization $\widetilde{\mathbf{P}}^{(\mathrm{lin})}\left(\mathbf{r},\omega\right) = \widetilde{\mathbf{P}}^{(1)}\left(\mathbf{r},\omega\right)$ yields

$$
\widetilde{\mathbf{P}}^{(1)}\left(\mathbf{r},\omega\right) = \epsilon_0\,\underline{\widetilde{\chi}}^{(1)}\left(\omega\right)\,\widetilde{\mathbf{E}}\left(\mathbf{r},\omega\right),
\tag{B.18}
$$

where the formal expression $\underline{\widetilde{\chi}}^{(1)}\left(\omega:\omega\right)$ was abbreviated by $\underline{\widetilde{\chi}}^{(1)}\left(\omega\right)$. For the linear optical susceptibility, no new frequencies are generated. We assume a superposition of narrowband signals centered at frequencies ω_ν, see Eq. (B.14). $\underline{\widetilde{\chi}}^{(1)}\left(\omega\right)$ can then be assumed constant over the bandwidth of each signal. The linear polarization can hence be written as

$$
\mathbf{P}^{(1)}\left(\mathbf{r},t\right) = \frac{1}{2}\sum_{\nu=-N_\Sigma}^{N_\Sigma}\left(1-\delta_{\nu,0}\right)\widehat{\underline{\mathbf{P}}}^{(1)}\left(\mathbf{r},t,\omega_\nu\right)\mathrm{e}^{\mathrm{j}\omega_\nu t},
\tag{B.19}
$$

where the slowly-varying polarization envelope $\widehat{\underline{\mathbf{P}}}^{(1)}\left(\mathbf{r},t,\omega_\nu\right)$ is given by

$$
\widehat{\underline{\mathbf{P}}}^{(1)}\left(\mathbf{r},t,\omega_\nu\right) = \epsilon_0\,\underline{\widetilde{\chi}}^{(1)}\left(\omega_\nu\right)\,\widehat{\underline{\mathbf{E}}}(\mathbf{r},t,\omega_\nu).
\tag{B.20}
$$

For a single narrowband signal centered at frequency $\omega_\mathrm{c}\neq 0$,

$$
\mathbf{E}(\mathbf{r},t) = \mathrm{Re}\left\{\widehat{\underline{\mathbf{E}}}(\mathbf{r},t,\omega_\mathrm{c})\,\mathrm{e}^{\mathrm{j}\omega_\mathrm{c}t}\right\},
\tag{B.21}
$$

the polarization and the electric field are related by a simple tensor product in the time domain,

$$
\mathbf{P}(\mathbf{r},t) = \epsilon_0\,\underline{\widetilde{\chi}}^{(1)}\left(\omega_c\right)\mathbf{E}(\mathbf{r},t).
\tag{B.22}
$$

B.3.2 Second-Order Nonlinear Polarization

To describe the nonlinear polarization by means of SVEA, we have to substitute the SVEA formulation, Eq. (B.14), in Eq. (B.7). The procedure is described in more detail for the slightly more complicated case of third-order nonlinear susceptibilities in the next section and the derivation for second-order processes is shortened. In the time-domain, the second-order nonlinear polarization reads

$$
\mathbf{P}^{(2)}(\mathbf{r},t) = \frac{\epsilon_0}{4}\sum_{\nu_1,\nu_2=-N}^{N}(1+\delta_{\nu_1,0})(1+\delta_{\nu_2,0})\,\underline{\widetilde{\chi}}^{(2)}(\omega_{\nu_1}+\omega_{\nu_2}:\omega_{\nu_1},\omega_{\nu_2}):\widehat{\underline{\mathbf{E}}}(\omega_{\nu_1})\widehat{\underline{\mathbf{E}}}(\omega_{\nu_2})\,\mathrm{e}^{\mathrm{j}\left(\omega_{\nu_1}+\omega_{\nu_2}\right)t},
\tag{B.23}
$$

where the arguments $(\mathbf{r}, t)$ of the electric fields $\underline{\widehat{\mathbf{E}}}$ have been omitted on the right-hand side. We would like to write the second-order nonlinear polarization in SVEA,

$$\mathbf{P}^{(2)}\left(\mathbf{r}, t\right) = \frac{1}{2} \sum_{\nu=-N_\Sigma}^{N_\Sigma} \left(1 - \delta_{\nu,0}\right) \underline{\widehat{\mathbf{P}}}^{(2)}\left(\mathbf{r}, t, \omega_{\Sigma\nu}\right) e^{j\omega_{\Sigma\nu}t}, \tag{B.24}$$

where the frequencies $\omega_{\Sigma\nu}$ represent all distinct frequencies that can be generated by second-order nonlinear interaction of the launched frequencies $\omega_{-N} \ldots \omega_N$,

$$\omega_{\Sigma\nu} \in \left\{\omega_{\nu_1} + \omega_{\nu_2} | \nu_{1,2} = -N \ldots N\right\}. \tag{B.25}$$

The labelling is chosen such that $\omega_{\Sigma-\nu} = -\omega_{\Sigma\nu}$ and, if there is a DC-component, $\omega_{\Sigma0} = 0$. The slowly varying envelope function $\underline{\widehat{\mathbf{P}}}^{(2)}\left(\mathbf{r}, t, \omega_{\Sigma\nu}\right)$ can thus be written as

$$\underline{\widehat{\mathbf{P}}}^{(2)}\left(\mathbf{r}, t, \omega_{\Sigma\nu}\right) = \frac{\epsilon_0}{2} \sum_{\mathbb{S}(\omega_{\Sigma\nu})} \frac{\left(1 + \delta_{\nu_1,0}\right)\left(1 + \delta_{\nu_2,0}\right)}{1 + \delta_{\nu,0}}$$
$$\times \underline{\widetilde{\chi}}^{(2)}\left(\omega_{\nu_1} + \omega_{\nu_2} : \omega_{\nu_1}, \omega_{\nu_2}\right) : \underline{\widehat{\mathbf{E}}}(\omega_{\nu_1})\underline{\widehat{\mathbf{E}}}(\omega_{\nu_2}) \, e^{j\left(\omega_{\nu_1} + \omega_{\nu_2}\right)t} \tag{B.26}$$

where

$$\mathbb{S}\left(\omega_{\Sigma\nu}\right) = \left\{(\nu_1, \nu_2) | \omega_{\Sigma\nu} = \omega_{\nu_1} + \omega_{\nu 2}, \, \nu_{1,2} = -N \ldots N\right\}. \tag{B.27}$$

Using the expansion in waveguide modes according to Eq. (B.16), the slowly varying envelope function $\underline{\widehat{\mathbf{P}}}^{(2)}\left(\mathbf{r}, t, \omega_{\Sigma\nu}\right)$ can be rewritten as

$$\underline{\widehat{\mathbf{P}}}^{(2)}\left(\mathbf{r}, t, \omega_{\Sigma\nu}\right) = \frac{\epsilon_0}{2} \sum_{\mathbb{S}(\omega_{\Sigma\nu})} \sum_{\mu_1,\mu_2} \frac{\left(1 + \delta_{\nu_1,0}\right)\left(1 + \delta_{\nu_2,0}\right)}{1 + \delta_{\nu,0}} \tag{B.28}$$
$$\times \underline{\widetilde{\chi}}^{(2)}\left(\omega_{\nu_1} + \omega_{\nu_2} : \omega_{\nu_1}, \omega_{\nu_2}\right) : \mathcal{E}_{\mu_1}\left(x, y, \omega_{\nu_1}\right) \mathcal{E}_{\mu_2}\left(x, y, \omega_{\nu_2}\right)$$
$$\times \frac{1}{\sqrt{\mathcal{P}_{\mu_1}\left(\omega_{\nu_1}\right) \mathcal{P}_{\mu_2}\left(\omega_{\nu_1}\right)}} A_{\mu_1}(z, t, \omega_{\nu_1}) A_{\mu_2}(z, t, \omega_{\nu_2}) \, e^{-\left(\beta_{\mu_1}\left(\omega_{\nu_1}\right) + \beta_{\mu_2}\left(\omega_{\nu_2}\right)\right)z}.$$

Refer to the next section for a more detailed description of the derivation.

B.3.3 Third-Order Nonlinear Polarization

Specializing Eq. (B.7) for a third-order nonlinear process yields

$$\underline{\widetilde{\mathbf{P}}}^{(3)}\left(\mathbf{r}, \omega\right) = \frac{\epsilon_0}{(2\pi)^2} \int\limits_{\omega_a} \int\limits_{\omega_b} d\omega_a \, d\omega_b \, \underline{\widetilde{\chi}}^{(3)}\left(\omega : \omega_a, \omega_b, \omega - \omega_a - \omega_b\right)$$
$$\vdots \, \underline{\widetilde{\mathbf{E}}}\left(\mathbf{r}, \omega_a\right) \underline{\widetilde{\mathbf{E}}}\left(\mathbf{r}, \omega_b\right) \underline{\widetilde{\mathbf{E}}}\left(\mathbf{r}, \omega - \omega_a - \omega_b\right). \tag{B.29}$$

By substituting the Fourier Transform of Eq. (B.14), we obtain

$$\widetilde{\mathbf{P}}^{(3)}\left(\mathbf{r},\omega\right) = \frac{\epsilon_0}{8 \times (2\pi)^2} \sum_{\nu_1,\nu_2,\nu_3=-N}^{N} \left(1+\delta_{\nu_1,0}\right)\left(1+\delta_{\nu_2,0}\right)\left(1+\delta_{\nu_3,0}\right) \tag{B.30}$$

$$\times \int\limits_{\omega_a}\int\limits_{\omega_b} \mathrm{d}\omega_a\,\mathrm{d}\omega_b\underline{\widetilde{\chi}}^{(3)}\left(\omega:\omega_a,\omega_b,\omega-\omega_a-\omega_b\right)$$

$$\vdots\ \underline{\widehat{\mathbf{E}}}(\mathbf{r},\omega_a-\omega_{\nu_1},\omega_{\nu_1})\underline{\widehat{\mathbf{E}}}(\mathbf{r},\omega_b-\omega_{\nu_2},\omega_{\nu_2})\underline{\widehat{\mathbf{E}}}(\mathbf{r},\omega-\omega_a-\omega_b-\omega_{\nu_3},\omega_{\nu_3})$$

where $\underline{\widehat{\mathbf{E}}}(\mathbf{r},\omega,\omega_{\nu_1})$ denotes the Fourier transform of $\underline{\widehat{\mathbf{E}}}(\mathbf{r},t,\omega_{\nu_1})$ with respect to t. The spectra $\underline{\widehat{\mathbf{E}}}(\mathbf{r},\omega_a-\omega_{\nu_1},\omega_{\nu_1})$ are narrow-band,

$$\widehat{\mathbf{E}}(\mathbf{r},\omega,\omega_{\nu_1}) \neq \mathbf{0} \qquad \text{only for} \qquad |\omega| < \Delta\omega_\nu, \tag{B.31}$$

where $\Delta\omega_\nu$ is the "bandwidth" of the ν-th envelope. The product of any vector components of $\underline{\widehat{\mathbf{E}}}(\mathbf{r},\omega_a-\omega_{\nu_1},\omega_{\nu_1})$, $\underline{\widehat{\mathbf{E}}}(\mathbf{r},\omega_b-\omega_{\nu_2},\omega_{\nu_2})$, and $\underline{\widehat{\mathbf{E}}}(\mathbf{r},\omega-\omega_a-\omega_b-\omega_{\nu_3},\omega_{\nu_3})$ is thus different from zero only for $\omega_a \approx \omega_{\nu_1} \wedge \omega_b \approx \omega_{\nu_2} \wedge \omega \approx \omega_{\nu_1}+\omega_{\nu_2}+\omega_{\nu_3}$. Over this frequency range, we may approximate the nonlinear optical susceptibility by its constant value determined by the carrier frequencies,

$$\underline{\widetilde{\chi}}^{(3)}\left(\omega:\omega_a,\omega_b,\omega-\omega_a-\omega_b\right) \approx \underline{\widetilde{\chi}}^{(3)}\left(\omega_{\nu_1}+\omega_{\nu_2}+\omega_{\nu_3}:\omega_{\nu_1},\omega_{\nu_2},\omega_{\nu_3}\right). \tag{B.32}$$

The optical susceptibility can thus be taken outside the integrals over ω_a and ω_b, and we can transform the relation back to the time domain. We obtain

$$\mathbf{P}^{(3)}\left(\mathbf{r},t\right) = \frac{\epsilon_0}{8} \sum_{\nu_1,\nu_2,\nu_3=-N}^{N} \left(1+\delta_{\nu_1,0}\right)\left(1+\delta_{\nu_2,0}\right)\left(1+\delta_{\nu_3,0}\right)$$

$$\times \underline{\widetilde{\chi}}^{(3)}\left(\omega_\Sigma:\omega_{\nu_1},\omega_{\nu_2},\omega_{\nu_3}\right)\ \vdots\ \underline{\widehat{\mathbf{E}}}(\mathbf{r},t,\omega_{\nu_1})\underline{\widehat{\mathbf{E}}}(\mathbf{r},t,\omega_{\nu_2})\underline{\widehat{\mathbf{E}}}(\mathbf{r},t,\omega_{\nu_3})$$

$$\times \mathrm{e}^{\mathrm{j}\left(\omega_{\nu_1}+\omega_{\nu_2}+\omega_{\nu_3}\right)t}. \tag{B.33}$$

The nonlinear polarization can be represented using slowly-varying complex amplitudes $\underline{\widehat{\mathbf{P}}}^{(3)}\left(\mathbf{r},t,\omega_{\Sigma\nu}\right)$,

$$\mathbf{P}^{(3)}\left(\mathbf{r},t\right) = \frac{1}{2} \sum_{\nu=-N_\Sigma}^{N_\Sigma} \left(1+\delta_{\nu,0}\right)\underline{\widehat{\mathbf{P}}}^{(3)}\left(\mathbf{r},t,\omega_{\Sigma\nu}\right)\mathrm{e}^{\mathrm{j}\omega_{\Sigma\nu}t}, \tag{B.34}$$

where the frequencies $\omega_{\Sigma\nu}$ represent all distinct frequencies that can be generated by third-order nonlinear interaction of the launched frequencies $\omega_{-N}\ldots\omega_N$,

$$\omega_{\Sigma\nu} \in \left\{\omega_{\nu_1}+\omega_{\nu_2}+\omega_{\nu_3}|\nu_{1,2,3}=-N\ldots N\right\}, \tag{B.35}$$

The labelling is chosen such that $\omega_{\Sigma-\nu} = -\omega_{\Sigma\nu}$ and, if there is a DC-component, $\omega_{\Sigma 0} = 0$. By comparing Eqs. (B.33) and (B.34), the slowly varying envelope function $\underline{\widehat{\mathbf{P}}}^{(3)}\left(\mathbf{r},t,\omega_{\Sigma\nu}\right)$ can be written as

$$\underline{\widehat{\mathbf{P}}}^{(3)}\left(\mathbf{r},t,\omega_{\Sigma\nu}\right) = \frac{\epsilon_0}{4} \sum_{\mathbb{S}(\omega_{\Sigma\nu})}\sum_{\mu_1,\mu_2,\mu_3}^{N} \frac{\left(1+\delta_{\nu_1,0}\right)\left(1+\delta_{\nu_2,0}\right)\left(1+\delta_{\nu_3,0}\right)}{1+\delta_{\nu,0}} \tag{B.36}$$

$$\times \underline{\widetilde{\chi}}^{(3)}\left(\omega_{\Sigma\nu}:\omega_{\nu_1},\omega_{\nu_2},\omega_{\nu_3}\right)\ \vdots\ \underline{\widehat{\mathbf{E}}}(\mathbf{r},t,\omega_{\nu_1})\underline{\widehat{\mathbf{E}}}(\mathbf{r},t,\omega_{\nu_2})\underline{\widehat{\mathbf{E}}}(\mathbf{r},t,\omega_{\nu_3})$$

where

$$\mathbb{S}\left(\omega_{\Sigma\nu}\right) = \left\{\left(\nu_1, \nu_2, \nu_3\right) \middle| \omega_{\Sigma\nu} = \omega_{\nu_1} + \omega_{\nu_2} + \omega_{\nu_3}, \, \nu_{1,2,3} = -N \ldots N\right\}. \tag{B.37}$$

The sum $\sum_{\mathbb{S}(\omega_{\Sigma\nu})}$ in Eq. (B.36) runs over all triples of frequency-subscripts (ν_1, ν_2, ν_3) that lead to the same sum frequency $\omega_{\Sigma\nu} = \omega_{\nu_1} + \omega_{\nu_2} + \omega_{\nu_3}$. Evaluating this sum involves two steps: First all distinct subsets of three frequencies that fulfill $\omega_{\Sigma\nu} = \omega_{\nu_1} + \omega_{\nu_2} + \omega_{\nu_3}$ have to be identified. Second, one has to take into account all distinct permutations of each of these subsets. Depending on the exact values of the frequencies $\omega_{\nu_1}, \omega_{\nu_2}$ and ω_{ν_3}, the set of distinct permutations can comprise one, three, or six elements. This leads to different factors of degeneracy for different nonlinear processes. Note that there are also different degeneracy factors depending on the number of zero carrier frequencies that are involved in the process.

Using the expansion in waveguide modes according to Eq. (B.16), The slowly varying envelope function $\widehat{\mathbf{P}}^{(3)}\left(\mathbf{r}, t, \omega_{\Sigma\nu}\right)$ can be rewritten as

$$\begin{aligned}
\widehat{\mathbf{P}}^{(3)}\left(\mathbf{r}, t, \omega_{\Sigma\nu}\right) = &\frac{\epsilon_0}{4} \sum_{\mathbb{S}(\omega_{\Sigma\nu})} \sum_{\mu_1, \mu_2, \mu_3}^{N} \frac{\left(1 + \delta_{\nu_1,0}\right)\left(1 + \delta_{\nu_2,0}\right)\left(1 + \delta_{\nu_3,0}\right)}{1 + \delta_{\nu,0}} \\
&\times \underline{\widetilde{\chi}}^{(3)}\left(\omega_{\Sigma\nu} : \omega_{\nu_1}, \omega_{\nu_2}, \omega_{\nu_3}\right) \, \vdots \, \mathcal{E}_{\mu_1}\left(x, y, \omega_{\nu_1}\right) \mathcal{E}_{\mu_2}\left(x, y, \omega_{\nu_2}\right) \mathcal{E}_{\mu_3}\left(x, y, \omega_{\nu_3}\right) \\
&\times \frac{1}{\sqrt{\mathcal{P}_{\mu_1}\left(\omega_{\nu_1}\right)\mathcal{P}_{\mu_2}\left(\omega_{\nu_2}\right)\mathcal{P}_{\mu_3}\left(\omega_{\nu_3}\right)}} A_{\mu_1}(z, t, \omega_{\nu_1}) A_{\mu_2}(z, t, \omega_{\nu_2}) A_{\mu_2}(z, t, \omega_{\nu_3}) \\
&\times e^{-j\left(\beta_{\mu_1}\left(\omega_{\nu_1}\right) + \beta_{\mu_2}\left(\omega_{\nu_2}\right) + \beta_{\mu_3}\left(\omega_{\nu_3}\right)\right)z}.
\end{aligned} \tag{B.38}$$

Appendix C

Signal Propagation and Mode Coupling

C.1 Linear Signal Propagation and Mode Coupling Equations

An optical impulse changes its shape when propagating along a waveguide. If the waveguide is ideal in the sense that it is completely uniform in the direction of propagation, an optical signal is only affected by waveguide dispersion effects. However, the waveguide can be affected by unintended perturbations of its structure such as sidewall roughness, or by intended modifications such as corrugations, or it can be subject to optical nonlinearities.

Using first-order perturbation theory, equations for the evolution of an optical signal shape can be derived also for such a "non-ideal" waveguide. The perturbation analysis remains valid as long as the modifications on the dielectric structure of the waveguide are sufficiently weak. The deviations from the ideal structure can be described by additional expressions $\underline{\mathbf{\Delta}}_1(\mathbf{r}, t)$ and $\underline{\mathbf{\Delta}}_2(\mathbf{r}, t)$ on the right-hand side of Maxwell's curl equations (1.2) and (1.1). If the material is nonmagnetic, $\underline{\mathbf{\Delta}}_2(\mathbf{r}, t)$ vanishes, but for the sake of generality, we will keep the expression during the derivation of the coupled-mode equations and set it to zero afterwards. For the complex (analytical) time-domain quantities, we obtain

$$\nabla \times \underline{\mathbf{E}}(\mathbf{r}, t) = -\mu \frac{\partial \underline{\mathbf{H}}(\mathbf{r}, t)}{\partial t} - \frac{\partial \underline{\mathbf{\Delta}}_H(\mathbf{r}, t)}{\partial t}, \tag{C.1}$$

$$\nabla \times \underline{\mathbf{H}}(\mathbf{r}, t) = \underline{\epsilon}(\mathbf{r}) \frac{\partial \underline{\mathbf{E}}(\mathbf{r}, t)}{\partial t} + \frac{\partial \underline{\mathbf{\Delta}}_E(\mathbf{r}, t)}{\partial t}, \tag{C.2}$$

where $\mu = \mu_0$, $\underline{\epsilon}(\mathbf{r}) = \epsilon_0 \underline{\epsilon}_r(\mathbf{r})$ and $\underline{\epsilon}_r(\mathbf{r})$ corresponds to the permeability tensor of the ideal waveguide profile. For isotropic media, a scalar $\epsilon_r(\mathbf{r})$ can be used instead of the tensor $\underline{\epsilon}_r(\mathbf{r})$.

We expand the electric and the magnetic fields in modes of the ideal structure and use a SVEA-representation of the amplitude to describe the effects of the waveguide perturbation. The waveguide modes can be guided or radiative, and for a single carrier frequency ω_c, our ansatz similar to Eqs. (1.35) and (1.36) reads:

$$\mathbf{\underline{E}}(\mathbf{r}, t) = \sum_{\zeta} \oint_{\beta_{c}} \mathrm{d}\beta_{c}\, A_{\zeta,\beta_{c}}(z, t, \omega_{c}) \frac{\mathcal{E}_{\zeta,\beta_{c}}(x, y, \omega_{c})}{\sqrt{\mathcal{P}_{\zeta,\beta_{c}}}}\, \mathrm{e}^{-\mathrm{j}(\omega_{c} t - \beta_{c} z)}\,, \tag{C.3}$$

$$\mathbf{\underline{H}}(\mathbf{r}, t) = \sum_{\zeta} \oint_{\beta_{c}} \mathrm{d}\beta_{c}\, A_{\zeta,\beta_{c}}(z, t, \omega_{c}) \frac{\mathcal{H}_{\zeta,\beta_{c}}(x, y, \omega_{c})}{\sqrt{\mathcal{P}_{\zeta,\beta_{c}}}}\, \mathrm{e}^{-\mathrm{j}(\omega_{c} t - \beta_{c} z)}\,, \tag{C.4}$$

and

$$\mathbf{\underline{\widetilde{E}}}(\mathbf{r}, \omega) = \sum_{\zeta} \oint_{\beta_{c}} \mathrm{d}\beta_{c}\, \widetilde{A}_{\zeta,\beta_{c}}(z, \omega - \omega_{c}, \omega_{c}) \frac{\mathcal{E}_{\zeta,\beta_{c}}(x, y, \omega_{c})}{\sqrt{\mathcal{P}_{\zeta,\beta_{c}}}}\, \mathrm{e}^{-\mathrm{j}\beta_{c} z}\,, \tag{C.5}$$

$$\mathbf{\underline{\widetilde{H}}}(\mathbf{r}, \omega) = \sum_{\zeta} \oint_{\beta_{c}} \mathrm{d}\beta_{c}\, \widetilde{A}_{\zeta,\beta_{c}}(z, \omega - \omega_{c}, \omega_{c}) \frac{\mathcal{H}_{\zeta,\beta_{c}}(x, y, \omega_{c})}{\sqrt{\mathcal{P}_{\zeta,\beta_{c}}}}\, \mathrm{e}^{-\mathrm{j}\beta_{c} z}\,, \tag{C.6}$$

where the symbol $\oint$ is to be understood as a sum over the discrete propagation constants $\beta_{c,\nu}$ of the guided modes plus an integral over the continuum of the propagation constants β_{c} that belong to the various radiation modes at the carrier frequency ω_{c}. $A_{\zeta,\beta_{c}}(z, t, \omega_{c})$ is the amplitude of the mode indexed by ζ and β_{c}. The ansatz introduced into the Eqs. (C.1) and (C.2), and using the identity $\nabla \times (\Phi \mathbf{F}) = \Phi (\nabla \times \mathbf{F}) + (\nabla \Phi) \times \mathbf{F}$ we obtain

$$\sum_{\zeta} \oint_{\beta_{c}} \mathrm{d}\beta_{c} \left[\left(\widetilde{A}_{\zeta,\beta_{c}}\, \mathrm{e}^{-\mathrm{j}\beta_{c} z} \right) \nabla \times \frac{\mathcal{E}_{\zeta,\beta_{c}}(x, y, \omega_{c})}{\sqrt{\mathcal{P}_{\zeta,\beta_{c}}}} + \frac{\partial}{\partial z} \left(\widetilde{A}_{\zeta,\beta_{c}}\, \mathrm{e}^{-\mathrm{j}\beta_{c} z} \right) \mathbf{e}_{z} \times \frac{\mathcal{E}_{\zeta,\beta_{c}}(x, y, \omega_{c})}{\sqrt{\mathcal{P}_{\zeta,\beta_{c}}}} \right]$$

$$= -\mathrm{j}\omega\mu \sum_{\zeta} \oint_{\beta_{c}} \mathrm{d}\beta_{c} \left[\widetilde{A}_{\zeta,\beta_{c}} \frac{\mathcal{H}_{\zeta,\beta}(x, y, \omega_{c})}{\sqrt{\mathcal{P}_{\zeta,\beta_{c}}}}\, \mathrm{e}^{-\mathrm{j}\beta_{c} z} \right] - \mathrm{j}\omega \mathbf{\underline{\Delta}}_{H}(\mathbf{r}, \omega) \tag{C.7}$$

and a similar equation for the magnetic field,

$$\sum_{\zeta} \oint_{\beta_{c}} \mathrm{d}\beta_{c} \left[\left(\widetilde{A}_{\zeta,\beta_{c}}\, \mathrm{e}^{-\mathrm{j}\beta_{c} z} \right) \nabla \times \frac{\mathcal{H}_{\zeta,\beta_{c}}(x, y, \omega_{c})}{\sqrt{\mathcal{P}_{\zeta,\beta_{c}}}} + \frac{\partial}{\partial z} \left(\widetilde{A}_{\zeta,\beta_{c}}\, \mathrm{e}^{-\mathrm{j}\beta_{c} z} \right) \mathbf{e}_{z} \times \frac{\mathcal{H}_{\zeta,\beta_{c}}(x, y, \omega_{c})}{\sqrt{\mathcal{P}_{\zeta,\beta_{c}}}} \right]$$

$$= \mathrm{j}\omega\epsilon \sum_{\zeta} \oint_{\beta_{c}} \mathrm{d}\beta_{c} \left[\widetilde{A}_{\zeta,\beta_{c}} \frac{\mathcal{E}_{\zeta,\beta_{c}}(x, y, \omega_{c})}{\sqrt{\mathcal{P}_{\zeta,\beta_{c}}}}\, \mathrm{e}^{-\mathrm{j}\beta_{c} z} \right] + \mathrm{j}\omega \mathbf{\underline{\Delta}}_{E}(\mathbf{r}, \omega), \tag{C.8}$$

where we have dropped the arguments of the envelope $\widetilde{A}_{\zeta,\beta_{c}}(z, \omega - \omega_{c}, \omega_{c})$ for better readability. Since we assumed $\widetilde{A}_{\zeta,\beta_{c}}$ to be a narrow-band envelope, the mode fields of the unperturbed waveguide structure can be assumed to be frequency-independent, $\mathcal{E}_{\zeta,\beta}(x, y, \omega) \approx \mathcal{E}_{\zeta,\beta_{c}}(x, y, \omega_{c})$ and $\mathcal{H}_{\zeta,\beta}(x, y, \omega) \approx \mathcal{H}_{\zeta,\beta_{c}}(x, y, \omega_{c})$. We now have to make use of the fact that the mode fields of the unperturbed structure have to obey Eqs. (C.1) and (C.2) for $\mathbf{\underline{\Delta}}_{H}$, $\mathbf{\underline{\Delta}}_{E} = 0$. This leads to

$$\nabla \times \mathcal{E}_{\zeta,\beta_{c}}(x, y, \omega) = \mathrm{j}\beta_{\zeta,\beta_{c}}(\omega)\, \mathbf{e}_{z} \times \mathcal{E}_{\zeta,\beta_{c}}(x, y, \omega) - \mathrm{j}\omega\mu \mathcal{H}_{\zeta,\beta_{c}}(x, y, \omega), \tag{C.9}$$

and

$$\nabla \times \mathcal{H}_{\zeta,\beta_c}(x,y,\omega) = j\,\beta_{\zeta,\beta_c}(\omega)\,\boldsymbol{e}_z \times \mathcal{H}_{\zeta,\beta_c}(x,y,\omega) + j\,\omega\epsilon\mathcal{E}_{\zeta,\beta_c}(x,y,\omega). \tag{C.10}$$

These equations can be substituted into Eqs. (C.7) and (C.8), and we obtain

$$\sum_{\zeta}\oint_{\beta_c} d\beta_c \left(\frac{\partial \widetilde{A}_{\zeta,\beta}}{\partial z} + j\left(\beta_{\zeta,\beta_c}(\omega) - \beta_c\right)\widetilde{A}_{\zeta,\beta_c}\right) e^{-j\beta_c z}\,\boldsymbol{e}_z \times \frac{\mathcal{E}_{\zeta,\beta_c}(x,y,\omega_c)}{\sqrt{\mathcal{P}_{\zeta,\beta_c}}} = -j\,\omega\underline{\widetilde{\Delta}}_H(\mathbf{r},\omega), \tag{C.11}$$

$$\sum_{\zeta}\oint_{\beta_c} d\beta_c \left(\frac{\partial \widetilde{A}_{\zeta,\beta_c}}{\partial z} + j\left(\beta_{\zeta,\beta_c}(\omega) - \beta_c\right)\widetilde{A}_{\zeta,\beta_c}\right) e^{-j\beta_c z}\,\boldsymbol{e}_z \times \frac{\mathcal{H}_{\zeta,\beta_c}(x,y,\omega_c)}{\sqrt{\mathcal{P}_{\zeta,\beta_c}}} = j\,\omega\underline{\widetilde{\Delta}}_E(\mathbf{r},\omega). \tag{C.12}$$

To project out the different mode amplitudes, the orthogonality relations Eqs. (1.37), (1.38) and (1.39) are used. We first dot-multiply both sides of Eqs. (C.11) and (C.12) with $\mathcal{H}^*_{\zeta',\beta'_c}$ and $\mathcal{E}^*_{\zeta',\beta'_c}$, respectively, and then subtract them. The result is integrated over the entire cross-section, and using the identity $(\mathbf{a} \times \mathbf{b})\cdot\mathbf{c} = \mathbf{a}\cdot(\mathbf{b} \times \mathbf{c})$, the orthogonality relations can be applied. We finally obtain

$$\left(\frac{\partial \widetilde{A}_{\zeta',\beta'_c}(z,\omega-\omega_c)}{\partial z} + j\left(\beta_{\zeta',\beta'_c}(\omega) - \beta'_c\right)\widetilde{A}_{\zeta',\beta'}(z,\omega-\omega_c)\right)e^{-j\beta'_c z} \tag{C.13}$$

$$= -j\frac{\omega}{4\sqrt{\mathcal{P}_{\zeta',\beta'_c}}}\iint\left(\underline{\widetilde{\Delta}}_H(\mathbf{r},\omega)\cdot\mathcal{H}^*_{\zeta',\beta'}(x,y,\omega_c) + \underline{\widetilde{\Delta}}_E(\mathbf{r},\omega)\cdot\mathcal{E}^*_{\zeta',\beta'}(x,y,\omega_c)\right)dx\,dy.$$

For narrow-band signals, the dispersion relation of the waveguide can be expanded into a Taylor series around ω_c,

$$\beta_{\zeta,\beta_c}(\omega) = \sum_{q=0}^{\infty}\frac{1}{q!}\beta^{(q)}_{\zeta,\beta_c}(\omega_c)\cdot(\omega-\omega_c)^q \tag{C.14}$$

where $\beta^{(q)}_{\zeta,\beta_c}(\omega_c) = \left.\frac{d^q\beta_\zeta}{d\omega^q}\right|_{\omega=\omega_c}$. Eq. (C.13) can be transformed to the time-domain,

$$\left(\frac{\partial A_{\zeta',\beta'_c}(z,t)}{\partial z} - \sum_{q=1}^{\infty}(-j)^{(q+1)}\frac{1}{q!}\beta^{(q)}_{\zeta',\beta'_c}(\omega_c)\frac{\partial^q A_{\zeta',\beta'_c}(z,t)}{\partial t^q}\right)e^{j(\omega_c t - \beta'_c z)} \tag{C.15}$$

$$= -\frac{\partial}{\partial t}\left[\frac{1}{4\sqrt{\mathcal{P}_{\zeta',\beta'_c}}}\iint\left(\underline{\widetilde{\Delta}}_H(\mathbf{r},t)\cdot\mathcal{H}^*_{\zeta',\beta'}(x,y,\omega_c) + \underline{\widetilde{\Delta}}_E(\mathbf{r},t)\cdot\mathcal{E}^*_{\zeta',\beta'}(x,y,\omega_c)\right)dx\,dy\right]$$

We will now discuss special cases of this general equation.

C.1.1 Ideal Linear Waveguide

For an ideal waveguide with zero perturbation ($\widetilde{\underline{\Delta}}_H$, $\widetilde{\underline{\Delta}}_E = 0$), the pulse shape is influenced by dispersion only. We obtain

$$\frac{\partial \widetilde{A}_{\zeta',\beta_c'}(z, \omega - \omega_c)}{\partial z} + \mathrm{j}\left(\beta_{\zeta',\beta_c'}(\omega) - \beta_c\right) \widetilde{A}_{\zeta',\beta_c'}(z, \omega - \omega_c) = 0. \tag{C.16}$$

or

$$\frac{\partial A_{\zeta',\beta_c'}(z, t)}{\partial z} = \sum_{q=1}^{\infty} (-\mathrm{j})^{(q+1)} \frac{1}{q!} \beta_{\zeta',\beta_c'}^{(q)}(\omega_c) \frac{\partial^q A_{\zeta',\beta_c'}(z, t)}{\partial t^q}$$

$$= -\beta_{\zeta',\beta_c'}^{(1)}(\omega_c) \frac{\partial A(z, t)}{\partial t} + \mathrm{j}\frac{1}{2}\beta_{\zeta',\beta_c'}^{(2)}(\omega_c) \frac{\partial^2 A(z, t)}{\partial t^2} + \frac{1}{6}\beta_{\zeta',\beta_c'}^{(3)}(\omega_c) \frac{\partial^3 A(z, t)}{\partial t^3} + \dots. \tag{C.17}$$

C.1.2 Perturbations of the Dielectric Profile

If the perturbations are caused by deviations $\Delta\underline{\epsilon}(\mathbf{r})$ from the ideal dielectric profile $\underline{\epsilon}(\mathbf{r})$, the quantities $\boldsymbol{\Delta}_1$ and $\boldsymbol{\Delta}_2$ are themselves proportional to the electromagnetic field amplitudes $\mathbf{H}$ and $\mathbf{E}$, respectively. This leads to a linear set of coupled partial differential equations for the mode envelopes and is therefore referred to as "linear coupling" between modes. For the perturbations, we have

$$\underline{\Delta}_H(\mathbf{r}, t) = \mathbf{0}, \tag{C.18}$$

$$\underline{\Delta}_E(\mathbf{r}, t) = \Delta\underline{\epsilon}(\mathbf{r})\underline{E}(\mathbf{r}, t), \tag{C.19}$$

and using Eqs. (C.3) and (C.4), this leads to

$$\underline{\Delta}_H(\mathbf{r}, t) = \mathbf{0}, \tag{C.20}$$

$$\underline{\Delta}_E(\mathbf{r}, t) = \Delta\underline{\epsilon}(\mathbf{r}) \sum_\zeta \oint_{\beta_c} \mathrm{d}\beta_c \, A_{\zeta,\beta_c}(z, t) \frac{\mathcal{E}_{\zeta,\beta_c}(x, y, \omega_c)}{\sqrt{\mathcal{P}_{\zeta,\beta_c}}} \, \mathrm{e}^{\mathrm{j}(\omega_c t - \beta_c z)}. \tag{C.21}$$

The evolution of the envelope is thus given by

$$\left(\frac{\partial A_{\zeta',\beta_c'}(z, t)}{\partial z} - \sum_{q=1}^{\infty} (-\mathrm{j})^{(q+1)} \frac{1}{q!} \beta_{\zeta',\beta_c'}^{(q)}(\omega_c) \frac{\partial^q A_{\zeta',\beta_c'}(z, t)}{\partial t^q}\right) \mathrm{e}^{\mathrm{j}(\omega_c t - \beta_c' z)} \tag{C.22}$$

$$= -\frac{\partial}{\partial t} \left[\frac{1}{4} \sum_\zeta \oint_{\beta_c} \mathrm{d}\beta_c \, \frac{A_{\zeta,\beta_c}(z, t)\,\mathrm{e}^{-\mathrm{j}(\omega_c t - \beta_c z)}}{\sqrt{\mathcal{P}_{\zeta',\beta_c'}\mathcal{P}_{\zeta,\beta_c}}} \iint (\Delta\underline{\epsilon}(\mathbf{r})\mathcal{E}_{\zeta,\beta_c}(x, y, \omega_c)) \cdot \mathcal{E}_{\zeta',\beta'}^*(x, y, \omega_c)\mathrm{d}x\,\mathrm{d}y\right]$$

Defining the coupling coefficient between mode (ζ', β_c') and mode (ζ, β_c),

$$K_{(\zeta',\beta_c')(\zeta,\beta_c)}(z) = \frac{\omega_c}{4\sqrt{\mathcal{P}_{\zeta',\beta_c'}\mathcal{P}_{\zeta,\beta_c}}} \iint \left(\Delta\underline{\epsilon}(\mathbf{r})\mathcal{E}_{\zeta,\beta_c}(x, y, \omega_c) \cdot \mathcal{E}_{\zeta',\beta'}^*(x, y, \omega_c)\right) \mathrm{d}x\,\mathrm{d}y, \tag{C.23}$$

and using the SVEA for the envelope,

$$\left| \frac{\partial}{\partial t} A_{\zeta,\beta_{\rm c}}(z,t) \right| \ll \left| \omega_{\rm c} A_{\zeta,\beta_{\rm c}}(z,t) \right| , \tag{C.24}$$

we obtain

$$\left(\frac{\partial A_{\zeta',\beta_{\rm c}'}(z,t)}{\partial z} - \sum_{q=1}^{\infty} (-{\rm j})^{(q+1)} \frac{1}{q!} \beta_{\zeta',\beta_{\rm c}'}^{(q)}(\omega_{\rm c}) \frac{\partial^q A_{\zeta',\beta_{\rm c}'}(z,t)}{\partial t^q} \right) {\rm e}^{{\rm j}(\omega_{\rm c}t-\beta_{\rm c}'z)}$$

$$= -{\rm j} \sum_{\zeta} \oint_{\beta_{\rm c}} {\rm d}\beta_{\rm c}\, K_{(\zeta',\beta_{\rm c}')(\zeta,\beta_{\rm c})}(z)\, A_{\zeta,\beta_{\rm c}}(z,t)\, {\rm e}^{-{\rm j}\beta_{\rm c}z} . \tag{C.25}$$

Very often, Eq. (C.25) is of interest for cases where the optical power is distributed over many modes, all having different propagation constants and group velocities. Under these circumstances, the intermodal (multimode) dispersion dominates and the intramodal (waveguide) dispersion may be neglected by assuming $\beta_{\zeta',\beta_{\rm c}'}^{(2)} = 0$, $\beta_{\zeta',\beta_{\rm c}'}^{(3)} = 0\ldots$.

C.2 Nonlinear Pulse Propagation and Mode Coupling Equations

If the waveguide materials are nonlinear, the electromagnetic field in one mode is generally not only influenced by the other modes, but also by signals at other carrier frequencies. We will only consider the nonlinear coupling among the guided modes of a waveguide. Nonlinear interaction with and among radiation modes can safely be neglected, since the corresponding power densities are usually orders of magnitude too small to be of any significance. The guided modes form a discrete set and the mode expansion Eq. (C.5) and (C.6) reduces to a simple sum over the mode index μ.

Since nonlinear processes involve interaction between different frequencies, we have to extend Eq. (C.3) to the case where the signal in each mode consists of different carriers at frequencies ω_ν modulated with amplitudes $A_\mu(z,t,\omega_\nu)$. We ask how the nonlinear interaction generates signals at new carrier frequencies or changes the envelopes of existing carrier frequencies. We have to be aware of the fact that the nonlinear interaction of the launched frequencies will produce "first-generation" signals centered at carrier frequencies that were not necessarily launched, and these signals will interact further and create "second-, third- ... generation" spectral components. Starting from a finite set of initial excitation frequencies, we may eventually end up with an infinite set of carrier frequencies. However, in practical cases, nonlinear interactions among "first-generation" signals are already very weak, and this holds all the more for interactions between "second-, third-...generation" signals. For practical applications, it is therefore sufficient to limit the consideration to the carrier frequencies $\omega_{\Sigma\nu}$ that are produced by direct nonlinear interaction of the launched frequencies ω_ν.

For a n-th order nonlinear process, the possible frequencies $\omega_{\Sigma\nu}$, $\nu = -N_\Sigma \ldots N_\Sigma$ are found by evaluating all sums of n frequencies ω_ν, $\nu = -N \ldots N$, as discussed in more detail in Section B.3. The numbering of the "sum" frequencies $\omega_{\Sigma\nu}$ is done keeping the convention $\omega_{-\nu} = -\omega_\nu$; $\omega_0 = 0$.

Adopting the conventions of Eq. (B.17), we obtain an ansatz for the electric and the magnetic field can,

$$\mathbf{E}(\mathbf{r},t) = \frac{1}{2} \sum_{\nu=-N_\Sigma}^{N_\Sigma} \sum_\mu (1 + \delta_{\nu,0}) A_\mu(z,t,\omega_{\Sigma\nu}) \frac{\mathcal{E}_\mu(x,y,\omega_{\Sigma\nu})}{\sqrt{\mathcal{P}_\mu(\omega_{\Sigma\nu})}} e^{j(\omega_{\Sigma\nu}t - \beta_\mu(\omega_{\Sigma\nu})z)}, \qquad (C.26)$$

$$\mathbf{H}(\mathbf{r},t) = \frac{1}{2} \sum_{\nu=-N_\Sigma}^{N_\Sigma} \sum_\mu (1 + \delta_{\nu,0}) A_\mu(z,t,\omega_{\Sigma\nu}) \frac{\mathcal{H}_\mu(x,y,\omega_\nu)}{\sqrt{\mathcal{P}_\mu(\omega_\nu)}} e^{j(\omega_{\Sigma\nu}t - \beta_\mu(\omega_{\Sigma\nu})z)}. \qquad (C.27)$$

Splitting up the optical polarization into a linear and a nonlinear part according to Eqs. (1.8) and (1.9), we can restate Eqs. (C.1) and (C.2) for real time-domain quantities

$$\nabla \times \mathbf{E}(\mathbf{r},t) = -\mu \frac{\partial \mathbf{H}(\mathbf{r},t)}{\partial t} - \frac{\partial \mathbf{\Delta}_H(\mathbf{r},t)}{\partial t}, \qquad (C.28)$$

$$\nabla \times \mathbf{H}(\mathbf{r},t) = \underline{\epsilon}(\mathbf{r}) \frac{\partial \mathbf{E}(\mathbf{r},t)}{\partial t} + \frac{\partial \mathbf{\Delta}_E(\mathbf{r},t)}{\partial t}, \qquad (C.29)$$

where the perturbation on the right-hand side is determined by the nonlinear polarization,

$$\mathbf{\Delta}_H(\mathbf{r},t) = \mathbf{0}, \qquad (C.30)$$

$$\mathbf{\Delta}_E(\mathbf{r},t) = \mathbf{P}^{(\mathrm{nl})}(\mathbf{r},t)$$

$$= \frac{1}{2} \sum_{\nu=-N_\Sigma}^{N_\Sigma} (1 - \delta_{\nu,0}) \widehat{\mathbf{P}}^{(\mathrm{nl})}(\mathbf{r},t,\omega_{\Sigma\nu}) e^{j\omega_{\Sigma\nu}t} \qquad (C.31)$$

The nonlinear polarization is determined by the launched signals according to Eqs. (B.24) and (B.34). We first insert these equations together with the ansatz for the electric and the magnetic field into Eqs. (C.28) and (C.29). The SVEA ensures that the spectra of different carriers do not overlap and are thus orthogonal. The mode amplitudes can finally be projected out using the same procedure as for the linear case. The evolution of the envelope then follows the relation

$$\left(\frac{\partial A_{\mu'}(z,t,\omega_{\Sigma\nu})}{\partial z} - \sum_{q=1}^\infty (-j)^{(q+1)} \frac{1}{q!} \beta_{\mu'}^{(q)}(\omega_{\Sigma\nu}) \frac{\partial^q A_{\mu'}(z,t,\omega_{\Sigma\nu})}{\partial t^q} \right) e^{j\left(\omega_{\Sigma\nu}t - \beta_{\mu'}(\omega_{\Sigma\nu})z\right)}$$

$$= -\frac{\partial}{\partial t}\left[\frac{1}{4\sqrt{\mathcal{P}_{\mu'}(\omega_{\Sigma\nu})}} \iint \widehat{\mathbf{P}}^{(\mathrm{nl})}(\mathbf{r},t,\omega_{\Sigma\nu}) e^{j\omega_{\Sigma\nu}t} \cdot \mathcal{E}_{\mu'}^*(x,y,\omega_\mathrm{c}) \, \mathrm{d}x \, \mathrm{d}y \right] \qquad (C.32)$$

C.2.1   Second-Order Nonlinearities

To describe nonlinear effects of different order, we have to use the SVEA expressions for $\widehat{\mathbf{P}}^{(\mathrm{nl})}(\mathbf{r},t,\omega_{\Sigma\nu})$ as obtained in Section B.3. The derivation is described in more detail for the slightly more complicated case of third-order nonlinearities, and the derivation for second-order processes in this section can be performed in the same way.

For second-order nonlinearities, we have to consider the frequencies $\omega_{\Sigma\nu} \in \{\omega_{\nu_1} + \omega_{\nu_2} | \nu_{1,2} = -N \ldots N\}$. Defining the set of index pairs $\mathbb{S}(\omega_{\Sigma\nu})$ for each sum frequency $\omega_{\Sigma\nu}$ as in Eq. (B.25), the nonlinear coupling between the different modes can be described by

$$\left(\frac{\partial A_{\mu'}(z, t, \omega_{\Sigma\nu})}{\partial z} - \sum_{q=1}^{\infty} (-\mathrm{j})^{(q+1)} \frac{1}{q!} \beta_{\mu'}^{(q)}(\omega_{\Sigma\nu}) \frac{\partial^q A_{\mu'}(z, t, \omega_{\Sigma\nu})}{\partial t^q} \right) \tag{C.33}$$

$$= -\mathrm{j} \sum_{\mu_1,\,\mu_2} \sum_{\mathbb{S}(\omega_{\Sigma\nu})} d_{\mu':\mu_1\mu_2}^{(2)}(\omega_{\Sigma\nu} : \omega_{\nu_1}, \omega_{\nu_2})\, A_{\mu_1}(z, t, \omega_{\nu_1})\, A_{\mu_2}(z, t, \omega_{\nu_2})\, \mathrm{e}^{-\mathrm{j}\Delta\beta_{\mu':\mu_1\mu_2}(\omega_\nu:\omega_{\nu_1},\omega_{\nu_2})z}$$

where

$$d_{\mu':\mu_1\mu_2}(\omega_\nu : \omega_{\nu_1}, \omega_{\nu_2}) = \frac{\omega_\nu \epsilon_0}{8} \frac{(1 + \delta_{\nu_1,0})(1 + \delta_{\nu_2,0})}{\sqrt{\mathcal{P}_{\mu'}(\omega_{\Sigma\nu})\,\mathcal{P}_{\mu_1}(\omega_{\nu_1})\,\mathcal{P}_{\mu_2}(\omega_{\nu_2})}} \tag{C.34}$$

$$\times \iint \left[\tilde{\chi}^{(2)}(\omega_\nu : \omega_{\nu_1}, \omega_{\nu_2}) \vdots \mathcal{E}_{\mu_1}(\omega_{\nu_1})\,\mathcal{E}_{\mu_2}(\omega_{\nu_2}) \right] \cdot \mathcal{E}_{\mu'}^*(\omega_{\Sigma\nu})\, \mathrm{d}x\, \mathrm{d}y \tag{C.35}$$

and

$$\Delta\beta_{\mu':\mu_1\mu_2}(\omega_{\Sigma\nu} : \omega_{\nu_1}, \omega_{\nu_2}) = \beta_{\mu_1}(\omega_{\nu_1}) + \beta_{\mu_2}(\omega_{\nu_2}) - \beta_{\mu'}(\omega_{\Sigma\nu}). \tag{C.36}$$

The spatial arguments (x, y) have been omitted in Eq. (C.34).

C.2.2 Third-Order Nonlinearities

For third-order nonlinearities, we have to consider the frequencies $\omega_{\Sigma\nu} \in \{\omega_{\nu_1} + \omega_{\nu_2} + \omega_{\nu_3} | \nu_{1,2,3} = -N \ldots N\}$. The nonlinear polarization is given by Eq. (B.38), where, according to Eq. (B.37) the set $\mathbb{S}(\omega_{\Sigma\nu})$ contains all triples of indices of launched frequencies whose sum is $\omega_{\Sigma\nu}$,

$$\mathbb{S}(\omega_{\Sigma\nu}) = \{(\nu_1, \nu_2, \nu_3) | \omega_{\Sigma\nu} = \omega_{\nu_1} + \omega_{\nu_2} + \omega_{\nu_3}, \nu_{1,2,3} = -N \ldots N\}. \tag{C.37}$$

The derivative with respect to time on the right-hand side of Eq. (C.32) can be approximated using the slowly-varying envelope condition,

$$\frac{\partial}{\partial t} \left[\widehat{\mathbf{P}}^{(\mathrm{nl})}(\mathbf{r}, t, \omega_{\Sigma\nu})\, \mathrm{e}^{\mathrm{j}\omega_{\Sigma\nu}t} \right] \simeq \mathrm{j}\omega_{\Sigma\nu} \widehat{\mathbf{P}}^{(3)}(\mathbf{r}, t, \omega_{\Sigma\nu})$$

This leads to a relation for the evolution of the mode envelope $A_{\mu'}(z, t, \omega_{\Sigma\nu})$,

$$\left(\frac{\partial A_{\mu'}(z, t, \omega_{\Sigma\nu})}{\partial z} - \sum_{q=1}^{\infty} (-\mathrm{j})^{(q+1)} \frac{1}{q!} \beta_{\mu'}^{(q)}(\omega_{\Sigma\nu}) \frac{\partial^q A_{\mu'}(z, t, \omega_{\Sigma\nu})}{\partial t^q} \right) \tag{C.38}$$

$$= -\mathrm{j} \frac{1}{3} \sum_{\mu_1,\,\mu_2,\,\mu_3} \sum_{\mathbb{S}_{\omega\nu}} \gamma_{\mu':\mu_1\mu_2\mu_3}^{(3)}(\omega_{\Sigma\nu} : \omega_{\nu_1}, \omega_{\nu_2}, \omega_{\nu_3})\, A_{\mu_1}(z, t, \omega_{\nu_1}) A_{\mu_2}(z, t, \omega_{\nu_2}) A_{\mu_3}(z, t, \omega_{\nu_3})$$

$$\times\, \mathrm{e}^{-\mathrm{j}\Delta\beta_{\mu':\mu_1\mu_2\mu_3}(\omega_{\Sigma\nu}:\omega_{\nu_1},\omega_{\nu_2},\omega_{\nu_3})z}$$

where

$$\gamma_{\mu':\mu_1\mu_2\mu_3}\left(\omega_{\Sigma\nu}:\omega_{\nu_1},\omega_{\nu_2},\omega_{\nu_3}\right) = \frac{3\omega_\nu\epsilon_0}{16}\frac{\left(1+\delta_{\nu_1,0}\right)\left(1+\delta_{\nu_2,0}\right)\left(1+\delta_{\nu_3,0}\right)}{\sqrt{\mathcal{P}_{\mu'}\left(\omega_{\Sigma\nu}\right)\mathcal{P}_{\mu_1}\left(\omega_{\nu_1}\right)\mathcal{P}_{\mu_2}\left(\omega_{\nu_2}\right)\mathcal{P}_{\mu_3}\left(\omega_{\nu_3}\right)}} \tag{C.39}$$

$$\times \iint \left[\underline{\widetilde{\chi}}^{(3)}\left(\omega_{\Sigma\nu}:\omega_{\nu_1},\omega_{\nu_2},\omega_{\nu_3}\right) \; \vdots \; \mathcal{E}_{\mu_1}\left(\omega_{\nu_1}\right)\mathcal{E}_{\mu_2}\left(\omega_{\nu_2}\right)\mathcal{E}_{\mu_3}\left(\omega_{\nu_3}\right)\right] \cdot \mathcal{E}_{\mu'}^*\left(\omega_{\Sigma\nu}\right)\,\mathrm{d}x\,\mathrm{d}y$$

and

$$\Delta\beta_{\mu':\mu_1\mu_2\mu_3}\left(\omega_{\Sigma\nu}:\omega_{\nu_1},\omega_{\nu_2},\omega_{\nu_3}\right) = \beta_{\mu_1}\left(\omega_{\nu_1}\right) + \beta_{\mu_2}\left(\omega_{\nu_2}\right) + \beta_{\mu_3}\left(\omega_{\nu_3}\right) - \beta_{\mu'}\left(\omega_{\Sigma\nu}\right). \tag{C.40}$$

The spatial arguments (x, y) have again been omitted in Eq. (C.39).

Appendix D

Semianalytical Calculation of Radiation Modes

D.1 Orthogonality and Normalization of the Radiation Modes

In order for the radiation modes to be of practical interest, they should all be orthogonal to each other and should also be normalized so that the spectral power density is either constant or is at least known. If we separate the ρ and ζ dependence out of the mode index ν, and thus write the mode fields for the νth radiation mode as $(\mathbf{E}_\nu, \mathbf{H}_\nu) \equiv (\mathbf{E}_{\zeta,\rho}, \mathbf{H}_{\zeta,\rho})$, then we would like the radiation modes to satisfy the orthonormality relation [72, Chapter 8.5]

$$\frac{1}{4} \int\limits_{-\infty}^{+\infty} \mathrm{d}x \int\limits_{-\infty}^{+\infty} \mathrm{d}y \left[\mathbf{E}_{\zeta,\rho} \times \mathbf{H}^*_{\zeta',\rho'} + \mathbf{E}^*_{\zeta',\rho'} \times \mathbf{H}_{\zeta,\rho} \right]_z = P_{\zeta,\rho}\, \delta_{\zeta,\zeta'}\, \delta(\rho - \rho') \qquad (\text{D.1})$$

The function $P_{\zeta,\rho}\, \delta(\rho - \rho')$ is the spectral power density for the ζth radiation mode, where the density coefficient $P_{\zeta,\rho}$ can be set equal to our earlier symbol P by appropriate normalization. In contrast to Eq. (1.38), the orthogonality relation for radiation modes is defined in terms of the radial propagation constant ρ. Since $\rho^2 = k^2 - \beta^2$, see the paragraph following Eq. (3.15), the power densities $P_{\zeta,\beta}\delta(\beta - \beta')$ and $P_{\zeta,\rho}\delta(\rho - \rho')$ are related by $\beta P_{\zeta,\beta}\delta(\beta - \beta') = -\rho P_{\zeta,\rho}\delta(\rho - \rho')$.

A useful result from Sammut [93] states that, if the mode is split into an "incident" part (i.e., the part that would exist if the waveguide was not there) and a "response" part (i.e., the outgoing field from the waveguide), then the total radiation mode has the same orthogonality properties as that of the incident field. Expressed mathematically, if we write the incident field as $(\mathbf{E}^I, \mathbf{H}^I)$ and the scattered part as $(\mathbf{E}^S, \mathbf{H}^S)$, then first term in the overlap integral (D.1) becomes

$$\begin{aligned}
I &= \frac{1}{4} \int_0^{2\pi} \mathrm{d}\theta \int_0^{\infty} r\mathrm{d}r \left[\mathbf{E}_{\zeta,\rho} \times \mathbf{H}^*_{\zeta',\rho'} \right]_z \\
&= \frac{1}{4} \int_0^{2\pi} \mathrm{d}\theta \int_0^{\infty} r\mathrm{d}r \left[\mathbf{E}^I_{\zeta,\rho} \times \mathbf{H}^{I*}_{\zeta',\rho'} \right]_z
\end{aligned} \qquad (\text{D.2})$$

We note that the second term in the overlap integral can be treated in an identical manner, with subscripts changed appropriately.

The relation (D.2) is very convenient, because we have already used precisely this partition when constructing the radiation modes, and this allows us to write down the power density function $P_\zeta(\rho)$ explicitly. From the earlier expansion (3.19), the z-components of the incident field are

$$\left[\mathbf{E}^I_{\zeta,\rho}\right]_z = P_N J_N(\rho r)\sin(N\theta + \varphi) \tag{D.3}$$

$$\left[\mathbf{H}^I_{\zeta,\rho}\right]_z = Q_N J_N(\rho r)\cos(N\theta + \varphi) \ . \tag{D.4}$$

The tangential components of the incident field follow from differentiation via equations (3.21):

$$\left[\mathbf{E}^I_{\zeta,\rho}\right]_r = j\,P_N \frac{\beta_\rho J'_N(\rho r)}{\rho}\sin(N\theta + \varphi) - j\,Q_N k Z_0 \frac{N J_N(\rho r)}{\rho^2 r}\sin(N\theta + \varphi) \tag{D.5a}$$

$$\left[\mathbf{E}^I_{\zeta,\rho}\right]_\theta = j\,P_N \frac{\beta_\nu N J_N(\rho r)}{\rho^2 r}\cos(N\theta + \varphi) - j\,Q_N k Z_0 \frac{J'_N(\rho r)}{\rho}\cos(N\theta + \varphi) \tag{D.5b}$$

$$\left[\mathbf{H}^I_{\zeta,\rho}\right]_r = -j\,P_N \frac{k N J_N(\rho r)}{Z_0 \rho^2 r}\cos(N\theta + \varphi) + j\,Q_N \frac{\beta_\nu J'_N(\rho r)}{\rho}\cos(N\theta + \varphi) \tag{D.5c}$$

$$\left[\mathbf{H}^I_{\zeta,\rho}\right]_\theta = j\,P_N \frac{k J'_N(\rho r)}{Z_0 \rho}\sin(N\theta + \varphi) - j\,Q_N \frac{\beta_\nu N J_N(\rho r)}{\rho^2 r}\sin(N\theta + \varphi) \tag{D.5d}$$

The terms with r in the denominator do not contribute to the integral in (D.2) and can be neglected. The integral is then

$$I = \frac{1}{4}\int_0^{2\pi} d\theta \int_0^\infty r\,dr \left[\left[\mathbf{E}^I_{\zeta,\rho}\right]_r \left[\mathbf{H}^{I*}_{\zeta',\rho'}\right]_\theta - \left[\mathbf{E}^I_{\zeta,\rho}\right]_\theta \left[\mathbf{H}^{I*}_{\zeta',\rho'}\right]_r \right] \tag{D.6}$$

$$= \frac{1}{4}\int_0^{2\pi} d\theta \int_0^\infty r\,dr \left[P_N P^*_{N'}\left(\frac{k\beta_\nu}{Z_0 \rho\rho'}\right) J'_N(\rho r)J'_{N'}(\rho' r) \times \sin(N\theta + \varphi)\sin(N'\theta + \varphi')\right.$$

$$\left. + Q_N Q^*_{N'}\left(\frac{k Z_0 \beta'_\rho}{\rho\rho'}\right) J'_N(\rho r)J'_{N'}(\rho' r) \times \cos(N\theta + \varphi)\cos(N'\theta + \varphi')\right]$$

We do the θ integral first, to find

$$I = \frac{\pi}{2}\int_0^\infty r\,dr \left[\frac{k}{\rho\rho'}\left(\frac{\beta_\nu}{Z_0} P_N P^*_N + \beta'_\nu Z_0 Q_N Q^*_N\right) \times J'_N(\rho r)J'_N(\rho' r)\xi_N \delta_{N,N'}\delta_{\varphi,\varphi'}\right] \tag{D.7}$$

where $\xi_0 = 2$ and $\xi_N = 1$ for $N \neq 0$. The remaining integral, over r, can be evaluated using the expression

$$\int_0^\infty J'_N(\rho r)J'_N(\rho' r)r\,dr = \frac{\delta(\rho - \rho')}{\rho} \ . \tag{D.8}$$

The first half of the overlap integral is then

$$I = \frac{\pi k \beta_\nu}{4\rho^3 Z_0}\left(|P_N|^2 + Z_0^2 |Q_N|^2\right)\xi_N \delta(\rho - \rho')\delta_{N,N'}\delta_{\varphi,\varphi'} \ . \tag{D.9}$$

The second half of the overlap integral in (D.1) yields an identical result. Thus, the radiation modes are orthogonal, with spectral power density coefficient

$$P_{\zeta,\rho} = \frac{\pi k \beta_\nu}{2\rho^3 Z_0} \left(|P_N|^2 + Z_0^2 |Q_N|^2 \right) \xi_N \,. \tag{D.10}$$

This is the same result as presented for the free-space radiation modes described in [97].

D.2 Matrix coefficients for radiation modes

To shorten notation, we introduce the element-by-element product $A .* B$ of two matrices A and B,

$$C = A .* B \quad \Longleftrightarrow \quad c_{m\ell} = a_{m\ell}\, b_{m\ell}, \tag{D.11}$$

where A and B must have the same size.

The elements of the $M \times M$ matrices E^{LA}, E^{LC}, H^{LB}, H^{LD}, E^{TA}, E^{TB}, E^{TC}, E^{TD}, H^{TA}, H^{TB}, H^{TC}, and H^{TD} are then given by

$$E^{LA} = \mathrm{J} .* \mathrm{S} \tag{D.12}$$

$$E^{LC} = \mathrm{H} .* \mathrm{S} \tag{D.13}$$

$$H^{LB} = \mathrm{J} .* \mathrm{C} \tag{D.14}$$

$$H^{LD} = \mathrm{H} .* \mathrm{C} \tag{D.15}$$

$$E^{TA} = \beta_\nu \left(\mathbb{J}' .* \mathrm{S} .* \mathrm{R} + \mathbb{J} .* \mathrm{C} .* \mathrm{T} \right) \tag{D.16}$$

$$E^{TB} = -k Z_0 \left(\mathbb{J} .* \mathrm{S} .* \mathrm{R} + \mathbb{J}' .* \mathrm{C} .* \mathrm{T} \right) \tag{D.17}$$

$$E^{TC} = \beta_\nu \left(\mathbb{H}' .* \mathrm{S} .* \mathrm{R} + \mathbb{H} .* \mathrm{C} .* \mathrm{T} \right) \tag{D.18}$$

$$E^{TD} = -k Z_0 \left(\mathbb{H} .* \mathrm{S} .* \mathrm{R} + \mathbb{H}' .* \mathrm{C} .* \mathrm{T} \right) \tag{D.19}$$

$$H^{TA} = -\frac{n_g^2 k}{Z_0} \left(\mathbb{J} .* \mathrm{C} .* \mathrm{R} - \mathbb{J}' .* \mathrm{S} .* \mathrm{T} \right) \tag{D.20}$$

$$H^{TB} = \beta_\nu \left(\mathbb{J}' .* \mathrm{C} .* \mathrm{R} - \mathbb{J} .* \mathrm{S} .* \mathrm{T} \right) \tag{D.21}$$

$$H^{TC} = -\frac{k}{Z_0} \left(\mathbb{H} .* \mathrm{C} .* \mathrm{R} - \mathbb{H}' .* \mathrm{S} .* \mathrm{T} \right) \tag{D.22}$$

$$H^{TD} = \beta_\nu \left(\mathbb{H}' .* \mathrm{C} .* \mathrm{R} - \mathbb{H} .* \mathrm{S} .* \mathrm{T} \right) \tag{D.23}$$

The elements of the various matrices are given by

$$\mathrm{J}_{m\ell} = J_\ell(\eta r_m) \,, \quad \mathrm{H}_{m\ell} = H_\ell^{(1)}(\eta r_m) \,, \tag{D.24}$$

$$\mathbb{J}_{m\ell} = \frac{\ell J_\ell(\eta r_m)}{\eta^2 r_m} \,, \quad \mathbb{H}_{m\ell} = \frac{\ell H_\ell^{(1)}(\eta r_m)}{\eta^2 r_m} \,, \tag{D.25}$$

$$\mathbb{J}'_{m\ell} = \frac{J_\ell'(\eta r_m)}{\eta} \,, \quad \mathbb{H}'_{m\ell} = \frac{H_\ell^{(1)'}(\eta r_m)}{\eta} \,, \tag{D.26}$$

and

$$\mathrm{S}_{m\ell} = \sin(\ell\theta_m + \varphi), \; \mathrm{R}_{m\ell} = R(\theta_m), \tag{D.27}$$

$$\mathrm{C}_{m\ell} = \cos(\ell\theta_m + \varphi), \; \mathrm{T}_{m\ell} = T(\theta_m) \,. \tag{D.28}$$

The matrix components E^{LP}, H^{LQ}, E^{TP}, E^{TQ}, H^{TP} and H^{TQ} are:

$$E^{LP} = \mathrm{J}_e \mathbin{.\!*} \mathrm{S} \tag{D.29}$$

$$H^{LQ} = \mathrm{J}_e \mathbin{.\!*} \mathrm{C} \tag{D.30}$$

$$E^{TP} = \beta_\nu \left(\mathbb{J}'_e \mathbin{.\!*} \mathrm{S} \mathbin{.\!*} \mathrm{R} + \mathrm{J}_e \mathbin{.\!*} \mathrm{C} \mathbin{.\!*} \mathrm{T} \right) \tag{D.31}$$

$$E^{TQ} = -kZ_0 \left(\mathbb{J}_e \mathbin{.\!*} \mathrm{S} \mathbin{.\!*} \mathrm{R} + \mathbb{J}'_e \mathbin{.\!*} \mathrm{C} \mathbin{.\!*} \mathrm{T} \right) \tag{D.32}$$

$$H^{TP} = -\frac{k}{Z_0} \left(\mathbb{J}_e \mathbin{.\!*} \mathrm{C} \mathbin{.\!*} \mathrm{R} - \mathbb{J}'_e \mathbin{.\!*} \mathrm{S} \mathbin{.\!*} \mathrm{T} \right) \tag{D.33}$$

$$H^{TQ} = \beta_\nu \left(\mathbb{J}'_e \mathbin{.\!*} \mathrm{C} \mathbin{.\!*} \mathrm{R} - \mathbb{J}_e \mathbin{.\!*} \mathrm{S} \mathbin{.\!*} \mathrm{T} \right) \tag{D.34}$$

where

$$[\mathrm{J}_e]_{m\ell} = J_\ell(\rho r_m) \, , \tag{D.35}$$

$$[\mathbb{J}_e]_{m\ell} = \frac{\ell J_\ell(\rho r_m)}{\rho^2 r_m} \, , \tag{D.36}$$

$$[\mathbb{J}'_e]_{m\ell} = \frac{J'_\ell(\eta r_m)}{\eta} \, . \tag{D.37}$$

Appendix E

Useful Formulae and Constants

For quick reference and without any systematic introduction, this section summarizes formulae and physical constants that have proven useful in everyday's work.

- Physical constants:

$$\epsilon_0 = 8.854188 \times 10^{-6} \ \mathrm{A\,s\,/\,V\,m} \tag{E.1}$$

$$\mu_0 = 1.256637 \times 10^{-6} \ \mathrm{V\,s\,/\,A\,m} \tag{E.2}$$

$$Z_0 = \sqrt{\frac{\mu_0}{\epsilon_0}} = 376.7303 \ \Omega \tag{E.3}$$

$$c = \frac{1}{\sqrt{\mu_0 \epsilon_0}} = 2.99792458 \times 10^8 \ \mathrm{m\,/\,s} \tag{E.4}$$

- For a plane wave travelling in the positive z-direction in a homogeneous material, the transverse components of $\mathbf{E}$ and $\mathbf{H}$ are related by

$$H_x = -\frac{E_y}{Z_0} \tag{E.5}$$

$$H_y = \frac{E_x}{Z_0}. \tag{E.6}$$

- Derivatives of Bessel functions:

$$\frac{\mathrm{d}\,J_\nu(z)}{\mathrm{d}\,z} = \frac{1}{2}\left(J_{\nu-1}(z) - J_{\nu+1}(z)\right) \tag{E.7}$$

$$\frac{\mathrm{d}\,H_\nu^{(1)}(z)}{\mathrm{d}\,z} = \frac{1}{2}\left(H_{\nu-1}^{(1)}(z) - H_{\nu+1}^{(1)}(z)\right) \tag{E.8}$$

- Generating function for $J_l(x)$:

$$e^{j\,x\sin\varphi} = \sum_{\nu=-\infty}^{+\infty} J_\nu(x)\,e^{j\,\nu\varphi} \tag{E.9}$$

- Free spectral range (FSR) of a resonator (round-trip length L_{rt}, effective group refractive index n_{eg}, group velocity $v_{\text{g}} = c/n_{\text{eg}}$):

$$\Delta\omega_{\text{FSR}} = \frac{2\pi c}{L_{\text{rt}} n_{\text{eg}}} = \frac{2\pi v_{\text{g}}}{L_{\text{rt}}} \tag{E.10}$$

$$\Delta f_{\text{FSR}} = \frac{c}{L_{\text{rt}} n_{\text{eg}}} = \frac{v_g}{L_{\text{rt}}} \tag{E.11}$$

$$\Delta\lambda_{\text{FSR}} = \frac{\lambda^2}{L_{\text{rt}} n_{\text{eg}}} \tag{E.12}$$

- Conversion of linear power attenuation coefficient α_1 (unit m^{-1}) to attenuation a (in dB) per length L:

$$\frac{a|_{\text{dB}}}{L} = \alpha_1 \times 10 \lg e = 4.343\,\alpha_1 \tag{E.13}$$

Glossary

Special Symbols

$u(t)$ real time-domain signal, Eq. (A.1)

$\underline{u}(t)$ analytic time-domain signal, Eq. (A.5)

$\widetilde{u}(\omega)$ spectrum of $u(t)$, Eq. (A.1)

$\underline{\widetilde{u}}(\omega)$ single-sided spectrum of analytic signal $\underline{u}(t)$, Eq. (A.9)

$\widehat{\underline{\mathbf{F}}}(\mathbf{r}, t, \omega_\nu)$ slowly-varying envelope at carrier frequency ω_ν, Eq. (1.27)

∇ nabla operator, Eq. (1.2)

$\mathfrak{F}_t$ Fourier transformation with respect to t, Eq. (A.1)

$\mathfrak{F}_\omega^{-1}$ inverse Fourier transformation with respect to ω, Eq. (A.2)

$\mathfrak{H}_t$ Hilbert transformation with respect to t, Eq. (A.6)

$\mathfrak{P}$ Cauchy principal value, Eq. (A.6)

Calligraphic Symbols

$\mathcal{E}(x, y)$ vectorial electric waveguide mode field, unit V / m, Eq. (1.33)

$\mathcal{H}(x, y)$ vectorial electric waveguide mode field, unit A / m, Eq. (1.34)

$\mathcal{P}$ power or (power density) associated with a numerical mode field, unit W, Eqs. (1.40), (1.41)

$\mathcal{U}$ voltage associated with a numerical mode field, unit V, Eq. (6.2)

Greek Symbols

α power attenuation coefficient, unit m^{-1}, Eq. (1.85)

α_2 TPA coefficient, unit m / W, Eq. (7.3)

β modal propagation constant, unit m^{-1}, Eq. (1.33)

β_μ modal propagation constant of the μ-th guided mode, unit m^{-1}, Eq. (1.35)

$\beta^{(n)}$ n-th derivative of β with respect to ω, $\beta^{(n)} = \frac{\mathrm{d}^n \beta}{\mathrm{d}\omega^n}$, unit s^n / m, Eq. (1.48)

$\Delta\beta$ small change of modal propagation constant, unit m^{-1}, Eq. (1.57)

γ	nonlinear waveguide parameter, unit $(\mathrm{W\,m})^{-1}$, Eq. (1.79)
Γ	field confinement factor, Eq. (1.61)
$\delta_{\mu,\mu'}$	Kronecker delta symbol, Eq. (1.37)
$\delta(x)$	Dirac delta function, Eq. (1.38)
ϵ_0	electric permittivity of vacuum, $\epsilon_0 = 8.85419 \times 10^{-6}\ \mathrm{A\,s}/(\mathrm{V\,m})$, Eq. (1.6)
$\underline{\epsilon}_r$	dielectric permeability tensor, Eq. (1.10)
$\Delta\underline{\epsilon}_r$	small perturbation of the dielectric permeability tensor, Eq. (1.69)
ϵ_r	scalar relative dielectric permeability, Eq. (1.11)
$\Delta\epsilon_r$	small change of scalar realtive dielectric permeability, Eq. (1.56)
ζ	discrete modal parameter, Eq. (1.35)
$\underline{\eta}$	impermeability tensor, Eq. (1.73)
η_{FWM}	FWM conversion efficiency, Eq. (1.90)
κ	contour curvature, unit m^{-1}, Eq. (4.1)
κ_1, κ_2	amplitude coupling coefficients of ring resonator coupling zones, Fig. (9.1)
λ	wavelength, unit m
μ_0	magnetic permeability of vacuum, $\mu_0 = 1.25664 \times 10^{-6}\ \mathrm{V\,s}/(\mathrm{A\,m})$, Eq. (1.5)
μ_1, μ_2	amplitude transmission coefficients of ring resonator coupling zones, Fig. (9.1)
ν	summation index
$\Pi^\times$	normalized phase shift, unit $(\mathrm{V\,mm})^{-1}$, Eq. (6.4)
ρ	amplitude roundtrip transmission coefficients of ring resonator, Fig. (9.1)
τ_a, τ_b	time constants for decay of refractive index perturbation, unit s, Eq. (8.5)
τ_1, τ_2	time constants for decay of material gain perturbation, unit s, Eq. (8.4)
τ_1, τ_2, τ_3	time delays in pump-probe setup, unit s,Eqs. (2.2) and (2.5)
τ_g	group delay, unit s, Eq. (1.50)
τ_{rep}	period of the impulse trains, unit s, Eq. (2.2)
φ	local angle between a contour curve and the z-axis, Eq. (4.1)
ϕ	optical phase shift, Eq. (2.14)
$\Delta\phi$	pump-induced optical phase shift, Eq. (2.26)
$\underline{\chi}^{(n)}$	n-th order time-domain Volterra kernel of polarization response; tensor of rank $n+1$, unit $\mathrm{s}^{-n}(\mathrm{m}/\mathrm{V})^{n-1}$, Eq. (1.7)
$\chi^{(n)}_{q_0:q_1 q_2 \ldots q_n}$	elements of $\underline{\chi}^{(n)}$, Eq. (B.6)
$\underline{\widetilde{\chi}}^{(n)}$	n-th order n-th order susceptibility; tensor of rank $n+1$, unit $(\mathrm{m}/\mathrm{V})^{n-1}$, Eq. (B.6)

$\widetilde{\chi}^{(n)}_{q_0:q_1 q_2 \ldots q_n}$ elements of $\widetilde{\underline{\chi}}^{(n)}$, Eq. (B.5)

ω angular frequency, unit s^{-1} Eq. (1.27)

ω_c angular carrier frequency, unit s^{-1}, Eq. (1.27)

$\Delta\omega_\mathrm{FWHM}$ FWHM width in terms of angular frequency, unit s^{-1}, Eq. (9.1)

$\omega_\mathrm{ref}, \omega_\mathrm{prb}$ frequency shift of reference and probe signal, Eqs. (2.2) and (2.4)

Latin Symbols

$a(z,t)$ amplitude of a signal travelling along a waveguide, unit $\sqrt{\mathrm{W}}$, Eq. (2.2)

$A(z,t)$ slowly-varying envelope of a signal travelling along a waveguide, unit $\sqrt{\mathrm{W}}$, Eq. (1.42)

A_eff effective area for third-order nonlinear interaction, unit m^2, Eq. (1.79)

c_ν mode amplitude, Eq. (3.3)

D correlation length of surface roughness, Eq. (3.37)

D_2 dispersion parameter, unit $\mathrm{ps}/(\mathrm{km\,nm})$, Eq. (1.53)

$\mathbf{D}$ electric displacement, unit $\mathrm{A\,s}/\mathrm{m}^2$, Eq. (1.3)

$\mathbf{B}$ magnetic flux density, unit $\mathrm{V\,s}/\mathrm{m}^2$, Eq. (1.4)

$\mathbf{e}_z$ unit-vector along the z-direction

$\mathbf{E}$ electric field, unit V/m, Eq. (1.2)

$\mathbf{E}_\mathrm{mw}$ electric microwave field, unit V/m, Eq. (1.66)

$\underline{\mathbf{E}}_\mathrm{opt}$ electric field at optical frequency, unit V/m, Eq. (1.67)

f frequency, unit Hz

f_gd group-delay related $3\,\mathrm{dB}$-bandwidth, unit Hz, Eq. (6.28)

f_{RC} RC-limited $3\,\mathrm{dB}$-bandwidth, unit Hz, Eq. (6.15)

FE   field enhancement factor, Eq. (9.1)

$\mathrm{FOM}_\mathrm{TPA}$ TPA figure of merit, Eq. (7.4)

g time-domain Green's function of an LTV system, Eq. (A.17); generic taper function, Eq. (4.8); two-dimensional exposure profile, Eq. (4.18)

Δg small change of the material gain, unit m^{-1}, Eq. (8.4)

h impulse response of an LTI system, Eq. (A.17); two-dimensional imaging kernel, Eq. (4.18)

$H(\omega)$ Heaviside function, Eq. (A.9)

$H_\nu^{(1)}$ ν-th order Hankel function, Eq. (3.19)

$\mathbf{H}$ magnetic field, unit A/m, Eq. (1.1)

j   imaginary unit, $\mathrm{j}^2 = -1$

J_ν ν-th order Bessel function, Eq. (3.18)

k, k_0 free-space wavenumber, unit m^{-1}

K mode coupling coefficient, unit m^{-1}, Eq. (1.65)

n	refractive index profile, Eq. (1.11)
n_0	unperturbed refractive index profile of a waveguide, Eq. (3.2)
n_2	nonlinear refractive index, unit m^2/W, Eq. (7.2)
n_{EO}	refractive index of electro-optic material, Eq. (6.1)
Δn	small refractive index change, Eq. (1.56)
P	power, unit W, Eq. (1.46)
r_{mn}	electro-optic coefficient, unit (m/V), Eq. (1.76)
$\mathbf{r}$	vector of Cartesian coordinates x, y, and z, unit m
$\mathbf{r}_{ic}, \mathbf{r}_{oc}$	inner/outer contour trajectory, Eqs. (4.1) and (4.9)
R	etch rate, unit nm/min, Eq. (5.5)
R_{uv}	cross-correlation, Eq. (A.12)
S	responsitivity of a photodetector, Eq. (2.7)
T	amplitude transmission factor for LTV system, Eq. (2.14)
S_{uv}	energy/power density, Eqs. (A.15) and (A.16)
$\mathbf{S}$	Poynting vector, unit W/m^2, Eq. (1.23)
t	time, unit s
t'	retarded time, unit s, Eq. (1.54)
u, v, w	material (e.g., crystallographic) coordinate system, Eq. (1.76)
v_g	group velocity, unit m/s, Eq. (1.51)
x	(horizontal) spatial coordinate, unit m, Fig. 1.1
y	(vertical) spatial coordinate, unit m, Fig. 1.1
z	(longitudinal) spatial coordinate, unit m, Fig. 1.1
Z_0	free-space wave impedance, $Z_0 = \sqrt{\mu_0/\epsilon_0} = 376.7303\,\Omega$

Acronyms

ASE	amplified spontaneous emission
BOX	buried oxide
cw	continuous wave
CAIBE	chemically-assisted ion beam etching
CMOS	complementary metal-oxide-semiconductor
DUT	device under test
FE	field enhancement factor
FDTD	finite-difference time-domain
FSR	free spectral range
FTTH	fiber to the home
FWHM	full width at half the maximum
FWM	four-wave mixing

EDFA	erbium-doped fibre amplifier
GVD	group-velocity dispersion
HIC	high index-contrast
IBE	ion beam etching
ICP-RIE	inductively-coupled plasma reactive ion etching
LTI	linear time-invariant
LTV	linear time-variant
LPCVD	low-pressure chemical vapor deposition
MBE	molecular beam epitaxy
MEMS	micro-electro-mechanical system
MZI	Mach-Zehnder interferometer
MMI	multi-mode interference
NLA	nonlinear absorption
OTN	Optical Transport Network (ITU-T G.709)
PBG	photonic band gap
PECVD	plasma-enhanced chemical vapor deposition
PhC	photonic crystal
PIC	photonic integrated circuits
PLC	planar lightwave circuit
RIE	reactive ion etching
RIBE	reactive ion beam etching
RMS	root mean square
SOI	silicon-on-insulator
SOA	semiconductor optical amplifier
SPM	self-phase modulation
SOA	semiconductor optical amplifier
SEM	scanning electron microscope
SVEA	slowly-varying envelope approximation
QD	quantum dot
TOD	third-order dispersion
TPA	two-photon absorption
ULSI	ultra-large-scale integration
VLSI	very-large-scale integration
XGM	cross-gain modulation
XPM	cross-phase modulation

Bibliography

[1] P. P. Absil, J. V. Hryniewicz, B. E. Little, P. S. Cho, R. A. Wilson, L. G. Joneckis, and P.-T. Ho. Wavelength conversion in GaAs micro-ring resonators. *Opt. Lett.*, 25(8):554–556, 2000.

[2] S. Adachi. Optical properties of $In_{1-x}Ga_xAs_yP_{1-y}$ alloys. *Phys. Rev. B*, 39(17):12612–12621, 1989.

[3] G. P. Agrawal. *Nonlinear Fiber Optics*. Academic Press, San Diego, third edition, 2001.

[4] T. Akiyama, M. Ekawa, M. Sugawara, K. Kawaguchi, H. Sudo, H. Kuwatsuka, H. Ebe, A. Kuramata, and Y. Arakawa. Quantum dots for semiconductor optical amplifiers. In *Optical Fiber Communication (OFC) Conference*, Anaheim (CA), USA, March 2005. OWM2.

[5] V. R. Almeida, Q. Xu, C. A. Barrios, and M. Lipson. Guiding and confining light in void nanostructure. *Opt. Lett.*, 29(11):1209–1211, 2004.

[6] Apollo Photonics, http://www.apollophoton.com/apollo/. *Apollo Photonics Solutions Suite APSS 2.1*, 2003.

[7] M. Asobe, I. Yokohama, T. Kaino, S. Tomaru, and T. Kurihara. Nonlinear absorption and refraction in an organic dye functionalized main chain polymer waveguide in the 1.5μm wavelength region. *Appl. Phys. Lett.*, 67(7):891–893, 1995.

[8] T. Baehr-Jones, M. Hochberg, G. Wang, R. Lawson, Y. Liao, Sullivan P. A., L. Dalton, A. K.-Y. Jen, and A. Scherer. Optical modulation and detection in slotted silicon waveguides. *Opt. Express*, 13(14):5216–5226, July 2005.

[9] C. A. Barrios. High-performance all-optical silicon microswitch. *Electron. Lett.*, 40(14), 2004.

[10] B. Bellini, J.-F. Larchanch, J.-P. Vilcot, D. Decoster, R. Becherelli, and A. d'Allesandro. Photonics devices based on preferential etching. *Appl. Opt.*, 44(33):7181–7186, November 2005.

[11] T. W. Berg and J. Mørk Mork. Saturation and noise properties of quantum-dot optical amplifiers. *IEEE J. Quantum Electron.*, 40(11):1527–1539, November 2004.

[12] A. K. Bhowmik and M. Thakur. Self-phase modulation in polydiacetylene single crystal measured at 720-1064nm. *Opt. Lett.*, 26(12):902–904, 2000.

[13] W. Bogaerts, R. Baets, P. Dumon, V. Wiaux, S. Beckx, D. Taillaert, B. Luyssaert, J. Van Campenhout, P. Bienstmann, and D. Van Thourhout. Nanophotonic waveguides in silicon-on-insulator fabricated with CMOS technology. *J. Lightwave Technol.*, 23(1):401–412, 2005.

[14] R. W. Boyd. *Nonlinear Optics.* Academic Press, San Diego, 2003.

[15] M. Bruel. Process for the production of thin semiconductor material films, December, 20 1994. United States Patent Number 5,374,564.

[16] T. Cardinal, K. A. Richardson, H. Shim, A. Schulte, R. Beatty, K. Le Foulgoc, C. Meneghini, J. F. Viens, and A. Villeneuve. Non-linear optical properties of chalcogenide glasses in the system As–S–Se. *J. Non-Cryst. Solids.*, 256&257:353–360, 1999.

[17] S.-H. Chang and A. Taflove. Finite-difference time-domain model of lasing action in a four-level two-electron atomic system. *Opt. Express*, 12(16):3827–3833, 2004.

[18] D. Chen, H. R. Fetterman, A. Chen, W. H. Steier, L. R. Dalton, W. Wang, and Y. Shi. Demonstration of 110 GHz electro-optical polymer modulators. *Appl. Phys. Lett.*, 70(25):3335–3337, June 1997.

[19] L. A. Coldren and S. W. Corzine. *Diode Lasers and Photonic Integrated Circuits.* John Wiley and Sons, New York, 1995.

[20] L. Dalton, B. Robinson, A. Jen, P. Ried, B. Eichinger, P. Sullivan, A. Akelaitis, D. Bale, M. Haller, J. Luo, S. Liu, Y. Liao, K. Firestone, N. Bhatambrekar, S. Bhattacharjee, J. Sinness, S. Hammond, N. Buker, R. Snoeberger, M. Lingwood, H. Rommel, J. Amend, S.-H. Jang, A. Chen, and W. Steier. Electro-optic coefficients of 500pm/v and beyond for organic materials. *Proc. SPIE, Linear and Nonlinear Optics of Organic Materials V.*, 5935:593502–1 – 593502–12, 2005.

[21] A. M. Darwish, E. P. Ippen, H. Q. Le, J. P. Donnelly, and S. H. Groves. Optimization of four-wave mixing conversion efficiency in the presence of nonlinear loss. *Appl. Phys. Lett.*, 69(6):737–739, Aug. 1996.

[22] M. Dinu, F. Quochi, and H. Garcia. Third-order nonlinearities in silicon at telecom wavelengths. *Appl. Phys. Lett.*, 82(18):2954–2956, 2003.

[23] E. Dulkeith, Y. A. Vlasov, X. Chen, N. C. Panoiu, and R. M. Osgood. Self-phase-modulation in submicron silicon-on-insulator photonic wires. *Opt. Express*, 14(12):5524–5534, 2006.

[24] P. Dumon, W. Bogaerts, V. Wiaux, J. Wouters, S. Beckx, J. Van Campenhout, D. Taillaert, B. Luyssaert, P. Bienstmann, D. Van Thourhout, and R. Baets. Low-loss SOI photonic wires and ring resonators fabricated with deep UV lithography. *IEEE Photon. Technol. Lett.*, 16(5):1328, 2004.

[25] P. Dumon, G. Roelkens, W. Bogaerts, D. Van Thourhout, J. Wouters, S. Beckx, P. Jaenen, and R. Baets. Basic photonic wire components in silicon-on-insulator. In *IEEE Intern. Conf. Group IV Photonics*, Antwerp, Belgium, 2005. WA1.

[26] S. Dupont, A. Beauriain, P. Miska, M. Zegaoui, J.-P Vilcot, H. W. Li, M. Constant, D. Decoster, and J. Chazelas. Low-loss InGasAsP/InP submicron optical waveguides fabricated by ICP etching. *Electron. Lett.*, 40:865–866, 2004.

[27] W. Freude. *Ausgewählte Kapitel aus der Hochfrequenztechnik — Teil 1 Rauschen.* Insitute of High-Frequency and Quantum Electronics, Karlsruhe, Germany, 2003. Lecture notes.

[28] R. S. Friberg and P. W. Smith. Nonlinear optical glasses for ultrafast optical switches. *IEEE J. Quantum Electron.*, 23(12):2089, 1987.

[29] M. Fujii and W. J. R Hoefer. Application of wavelet-galerkin method to electrically large optical waveguide problems. *IEEE MTT-S Int. Microwave Symposium Digest*, TU3E-3:239–242, 2000.

[30] M. Fujii, C. Koos, C. Poulton, J. Leuthold, and W. Freude. Nonlinear FDTD analysis and experimental validation of four-wave mixing in InGaAsP/InP racetrack micro-resonators. *IEEE Photon. Technol. Lett.*, 18(2):361–363, February 2006.

[31] M. Fujii, C. Koos, C. Poulton, C. Sakagami, J. Leuthold, and W. Freude. A simple and rigorous verification technique for nonlinear FDTD algorithm by optical parametric four-wave mixing. *Microwave and Optical Technol. Lett.*, 48(1):88–91, Jan. 2006.

[32] M. Fujii, D. Lukashevich, I. Sakagami, and P. Russer. Convergence of FDTD and wavelet-collocation modeling of curved dielectric interface with the effective dielectric constant technique. *IEEE Microwave and Wireless Components Lett.*, 13(11):469–471, Nov. 2003.

[33] M. Fujii, M. Tahara, I. Sakagami, W. Freude, and P. Russer. High-order FDTD and auxiliary differential equation formulation of optical pulse propagation in 2D Kerr and Raman nonlinear dispersive media. *IEEE J. Quantum Electron.*, 40(2):175–182, February 2004.

[34] T. Fujisawa and M. Koshiba. All-optical logic gates based on nonlinear slot-waveguide couplers. *J. Opt. Soc. Am. B*, 23(4):684–691, 2006.

[35] T. Fujisawa and M. Koshiba. Guided modes of nonlinear slot waveguides. *IEEE Photon. Technol. Lett.*, 18(14):1530–1532, 2006.

[36] T. Fukazawa, F. Ohno, and T. Baba. Very compact arrayed waveguide grating using Si photonic wire waveguides. *Jpn. J. Appl. Phys.*, 43(5B):L673–L675, 2004.

[37] H. Fukuda, K. Yamada, T. Shoji, M. Takahashi, T. Tsuchizawa, T. Watanabe, J. Takahasi, and S. Itabashi. Four-wave mixing in silicon wire waveguides. *Opt. Express*, 13(12):4629–4637, June 2005.

[38] O J. Glembocki, D. D. Palik, G. R. de Guel, and D. L. Kendall. Hydration model for the molarity dependence of the etch rate of Si in aqueous alkali hydroxides. *J. Electrochem. Soc.*, 138(4):1055–1063, April 1991.

[39] J. E. Goell. A circular-harmonic computer analysis of rectangular waveguide modes. *Bell. Sys. Tech. J.*, 48:2133–2160, 1969.

[40] P. M. Goorjian and Y. Silberberg. Numerical simulations of light bullets using the full-vector time-dependent nonlinear Maxwell equations. *J. Opt. Soc. Am. B*, 14(11):3253–3260, 1997.

[41] F. Grillot, L. Vivien, S. Laval, D. Pascal, and E. Cassan. Size influence on the propagation loss induced by sidewall roughness in ultrasmall SOI waveguides. *IEEE Photon. Technol. Lett.*, 16:1661–1663, 2004.

[42] L. Gu, W. Jiang, X. Chen, L. Wang, and R. T. Chen. High speed silicon photonic crystal waveguide modulator for low voltage operation. *Appl. Phys. Lett.*, 90:0711105-1–071105-3, February 2007.

[43] U. Gubler. *Third-order nonlinear optical effects in organic materials.* PhD thesis, Swiss Federal Institute of Technology Zürich, 2000.

[44] U. Gubler and C. Bosshard. Molecular design for third-order nonlinear optics. *Advances in Polymer Science*, 158:123–191, 2002.

[45] B. W. Hakki and T. L. Paoli. Gain spectra in GaAs double-heterostructure injection lasers. *Journ. of Appl. Phys.*, 46:1299–1476, 1975.

[46] K. L. Hall, G. Lenz, E. P. Ippen, and G. Raybon. Heterodyne pump-probe technique for time-domain studies of optical nonlinearities in waveguides. *Opt. Lett.*, 17(12), 1992.

[47] J. M. Harbold, F. Ö. Ilday, F. W. Wise, J. S. Sanghera, V. Q. Nguyen, L. B. Shaw, and I. D. Aggarwal. Higly nonlinear As-S-Se glasses for all-optical switching. *Opt. Lett.*, 27(2):119–121, 2002.

[48] C. H. Henry. Theory of the linewidth of semiconductor lasers. *IEEE J. Quantum Electron.*, QE-18(2):259–264, February 1982.

[49] J. D. Jackson. *Classical Electrodynamics.* John Wiley and Sons, New York, 1998.

[50] S. G Johnson, M. L. Povinelli, M. Soljacic, A. Karalis, S. Jacobs, and J. D. Joannopolous. Roughness losses and volume-current methods in photonic-crystal waveguides. *Appl. Phys. B*, 81:283293, 2005.

[51] D. S. Jones. *The theory of generalised functions.* Cambridge University Press, Cambridge, 1980.

[52] T. Kaino. Waveguide fabrication using organic nonlinear optical materials. *J. Opt. A: Pure Appl. Opt.*, 2:R1–R7, 2000.

[53] M. Kappelt and D. Bimberg. Wet chemical etching of high quality V-grooves with {111}A sidewalls in (001) InP. *J. Electrochem. Soc.*, 143(10):3271–3273, October 1996.

[54] D. L. Kendall. Vertical etching of silicon at very high aspect ratios. *Annual Review of Materials Science*, 9:373–403, 1979.

[55] D. L. Kendall. A new theory for the anisotropic etching of silicon and some underdeveloped chemical micromachining concepts. *J. Vac. Sci. Technol. A*, 8(4):3598–3605, July 1990.

[56] K. Kikuchi, K. Taira, and N. Sugimoto. Highly nonlinear bismuth oxide-based glass fibers for all-optical signal processing. *Electron. Lett.*, 38(4):166, 2002.

[57] D. Y. Kim, M. Sundheimer, A. Otomo, G. Stegeman, W. H. G. Horsthuis, and G. R. Möhlmann. Third order nonlinearity of 4-dialkyamino-4'nitro-stilbene waveguides at 1319nm. *Appl. Phys. Lett.*, 63(3):290–292, 1993.

[58] G. T. A. Kovacs, N. I. Maluf, and K. P. Petersen. Bulk micromachining of silicon. *Proc. IEEE*, 86(8):1536–1551, August 1998.

[59] A. R. Kovsh, N. A. Maleev, A. E. Zhukov, S. S. Mikhrin, A. P. Vasil'ev, E. A. Semenova, Yu. M. Shernyakov, M. V. Maximov, D. A. Livshits, V. M. Usitonov, N. N. Ledentsov, D. Bimberg, and Zh. I. Alferov. InAs/InGaAs/GaAs quantum dot lasers of $1.3\mu m$ range with enhanced optical gain. *Journal of Crystal Growth*, 251:729–736, 2003.

[60] M. Kuznetsov and H. A. Haus. Radiation loss in dielectric waveguide structures by the volume current method. *IEEE J. Quantum Electron.*, 19:1505–1514, 1983.

[61] M. Lämmlin. *GaAs-Based Semiconductor Optical Amplifiers with Quantum Dots as an Active Medium*. PhD thesis, Technische Universität Berlin, 2006.

[62] B. L. Lawrence, M. Cha, J. U. Kang, W. Toruellas, G. Stegemann, G. Baker, J. Meth, and S. Etemad. Large purely refractive nonlinear index of single-crystal P-toluene sulphonate (PTS) at 1600nm. *Electron. Lett.*, 30(5):447–448, 1994.

[63] K. K. Lee, D. R. Lim, L. C. Kimerling, J. Shin, and F. Cerrina. Fabrication of ultralow-loss Si/SiO$_2$ waveguides by roughness reduction. *Opt. Lett.*, 26(23):1888–1890, 2001.

[64] K. K. Lee, D. R. Lim, H.-C. Luan, A. Agarwal, J. Foresi, and L. C. Kimerling. Effect of size and roughness on light-transmission in a Si/SiO$_2$ waveguide: Experiments and model. *Appl. Phys. Lett.*, 77:1617–1619, 2000.

[65] D. Lenz, D. Erni, and W. Bächtold. Modal loss coefficients for highly overmoded rectangular dielectric waceguides based on free space modes. *Opt. Express*, 12:1150–1156, 2004.

[66] J. Y. Y. Leong, P. Petropoulos, S. Asimakis, H. Ebendorff-Heideprim, R. C. Moore, Ken. Frampton, V. Finazzi, X. Feng, J. H. V. Price, T. M. Monro, and D. J. Richardson. A lead silicate holey fiber with $\gamma = 1860\,(\text{Wkm})^{-1}$ at 1550 nm. In *Optical Fiber Communication (OFC) Conference*, Anaheim (CA), USA, March 2005. PDP22.

[67] M. Lipson. Guiding, modulating, and emitting light on silicon — challenges and opportunities. *J. Lightwave Technol.*, 23(12):4222–4238, 2005.

[68] B. E. Little, J. S. Foresi, G. Steinmeyer, E. R. Toen, S. T. Chu, H. A. Haus, E. P. Ippen, L. C. Kimerling, and W. Greene. Ultra-compact Si–SiO$_2$ microring resonator optical channel dropping filters. *IEEE Photon. Technol. Lett.*, 10(4):549–551, 1998.

[69] A. Liu, L. Liao, D. Rubin, H. Nguyen, B. Cifticioglu, Y. Chetrit, N. Izhaky, and M. Paniccia. High-speed optical modulation based on carrier depletion in a silicon waveguide. *Opt. Express*, 15(2):660–668, January 2007.

[70] D. Marcuse. Mode conversion caused by surface imperfections of a dielectric slab waveguide. *Bell. Sys. Tech. J.*, 48:3187–3215, 1969.

[71] D. Marcuse. Radiation losses of the dominant mode in round dielectric waveguides. *Bell. Sys. Tech. J.*, 49:1665–1693, 1970.

[72] D. Marcuse. *Light Transmission Optics*. Van Nostrand Reinhold, New York, 1972.

[73] D. Marcuse. Power distribution and radiation losses in multimode dielectric waveguides. *Bell. Sys. Tech. J.*, 51:429–454, 1972.

[74] D. Marcuse. *Theory of Dielectric Optical Waveguides*. Academic Press, London, 1974.

[75] W. Menz, J. Mohr, and O. Paul. *Microsystem Technology*. Wiley-VCH, New York, 3 edition, 2001.

[76] S. Mikroulis, H. Simos, E. Roditi, and D. Syvridis. Ultrafast all-optical AND logic operation based on four-wave mixing in a passive InGaAsP–InP microring resonator. *IEEE Photon. Technol. Lett.*, 17(9):1878–1880, September 2005.

[77] V. Mizrahi, K. W. DeLong, G. I. Stegeman, M. A. Saifi, and M. J. Andrejco. Two-photon absorption as a limitation to all-optical switching. *Opt. Lett.*, 14(20):1140–1142, December 1989.

[78] F. J. Morin and J. P. Maita. Electrical properties of silicon containing arsenic and boron. *Phys. Rev.*, 96(1):28–35, October 1954.

[79] P. Müllner and R. Hainberger. Structural optimization of silicon-on-insulator slot waveguides. *IEEE Photon. Technol. Lett.*, 18(24):2557–2559, 2006.

[80] H. Nasu, O. Matsushita, K. Kamiya, H. Kobayashi, and K. Kubodera. Third harmonic generation from Li$_2$O–TiO$_2$–TeO$_2$ glasses. *J. Non-Cryst. Solids.*, 124:275–277, 1990.

[81] K. Okamoto. *Fundamentals of Optical Waveguides.* Academic Press, San Diego, 2000.

[82] E. D. Palik, H. F. Gray, and P. B. Klein. A Raman study of etching silicon in aqueous KOH. *J. Electrochem. Soc.*, 130(4):956–959, April 1983.

[83] A. Papoulis. *Probability, random variables and stochastic processes*, pages 319–327. McGraw Hill, New York, 1991.

[84] F. P Payne and J. P. R. Lacey. A theoretical analysis of scattering loss from planar optical waveguides. *IEEE Proc. Optical and Quantum Electron.*, 26:997–986, 1994.

[85] K. E. Petersen. Silicon as a mechnical material. *Proc. IEEE*, 70(5):420–457, May 1982.

[86] C. G. Poulton, C. Koos, M. Fujii, A. Pfrang, Th. Schimmel, J. Leuthold, and W. Freude. Radiation modes and roughness loss in high index-contrast waveguides. *IEEE J. Sel. Top. Quantum Electron.*, 16(6), November/December 2006.

[87] C. M. Reinke, A. Jafarpour, B. Momeni, M. Soltani, S. Khorasani, A. Adibi, and R. K. Lee. Nonlinear finite-difference time-domain method for the simulation of anisotropic, $\chi^{(2)}$, and $\chi^{(3)}$ optical effects. *IEEE J. Lightwave Technol.*, 24(1):624–634, Jan. 2006.

[88] A. Reisman, M. Berkenblit, S. A. Chan, F. B. Kaufman, and D. C. Green. The controlled etching of silicon in catalyzed ethlyenediamine-pyrocatechol-water solutions. *J. Electrochem. Soc.*, 126(8):1406–1415, 1979.

[89] K. Rochford, R. Zanoni, G. I. Stegeman, W. Krug, E. Miao, and M. W. Beranek. Measurement of nonlinear refractive index and transmission in polydiacetlyene waveguides at 1.319μm. *Appl. Phys. Lett.*, 58(1):13–15, 1991.

[90] RSoft Design Group, Inc., http://www.rsoftdesign.com. *BeamPROP 6.0 User Guide*, 2005.

[91] RSoft Design Group, Inc., http://www.rsoftdesign.com. *FemSIM 1.0 User Guide*, 2005.

[92] RSoft Design Group, Inc., http://www.rsoftdesign.com. *FemSIM 2.0 User Guide*, 2005.

[93] R. A. Sammut. Orthogonality and normalisation of radiation modes in dielectric waveguides. *J. Opt. Soc. Am.*, 72:1335–1337, 1982.

[94] H. Seidel, L. Cspergi, A. Heuberger, and H. Baumgärtel. Anisotropic etching of crystalline silicon in alkaline solutions. *J. Electrochem. Soc.*, 137(11), November 1990.

[95] Y. R. Shen. *Nonlinear Optics.* John Wiley and Sons, New York, 1984.

[96] G. M. Slavcheva, J. M. Arnold, and R. W. Ziolkowski. FDTD simulation of the non-linear gain dynamics in active optical waveguides and semiconductor microcavities. *IEEE J. Sel. Top. Quantum Electron.*, 10(5):1052–1062, 2004.

[97] A. W. Snyder and J. D. Love. *Optical Waveguide Theory*, pages 514–533. Chapman & Hall, London, 1983.

[98] A. W. Snyder and R. A. Sammut. Radiation modes of optical waveguides. *Electron. Lett.*, 15:4–5, 1979.

[99] R. A. Soref and J. P. Lorenzo. All-silicon active and passive guided-wave components for $\lambda = 1.3\,\mu$m and $1.6\,\mu$m. *IEEE J. Quantum Electron.*, QE-22(6):873–879, June 1986.

[100] E. Steinsland, T. Finstad, and A. Hanneborg. Etch rates of (100), (111) and (110) single-crystal silicon in TMAH measured in situ by laser reflectance interferometry. *Sensors and Actuators*, 86:73–80, 2000.

[101] J. A. Stratton. *Electromagnetic theory*. McGraw-Hill, London, 1941.

[102] A. S. Sudbo. Why are accurate computations of mode fields in rectangular waveguides so difficult? *J. Lightwave Technol.*, 10:418–419, 1992.

[103] M. Sugawara, T. Akiyama, N. Hatori, Y. Nakata, H. Ebe, and H. Ishikawa. Quantum-dot semiconductor optical amplfiers-for-high-bit-rate signal processing up to $160\,\mathrm{Gb\,s^{-1}}$ and a new scheme of 3R regenerators. *Meas. Sci. Technol.* , 13:1683–1691, 2002.

[104] M. Sugawara, H. Ebe, N. Hatori, M. Ishida, Y. Arakawa, T. Akiyama, K. Otsubo, and Y. Nakata. Theory of optical gain amplification and processing by qunatum-dot semiconductor optical amplifiers. *Phys. Rev. B*, 69:235332–1–235332–39, 2004.

[105] O. Tabata, R. Asahi, H. Funabashi, K. Shimaoka, and S. Sugiyama. Anisotropic etching of silicon in TMAH solutions. *Sensors and Actuators A*, 34(1):51–57, July 1992.

[106] A. Taflove and S. C. Hagness. *Computational electrodynamics: The finite-difference time-domain method, 3rd ed.* Artech House, 2005.

[107] Y. Tarui, Y. Komiya, and Y. Harada. Preferential etching and etched profile of GaAs. *J. Electrochem. Soc.*, 118(1):118–122, 1971.

[108] P. K. Tien. Light waves in thin films and integrated optics. *AO*, 10:2395–2413, 1971.

[109] P. D. Townsend, G. L. Baker, N. E. Schlotter, C. F. Klausner, and S. Eternad. Waveguiding in spun films of soluble polydiacetylenes. *Appl. Phys. Lett.*, 53(19):1782–1784, 1988.

[110] H. K. Tsang, C. S. Wong, T. K. Liang, I. E. Day, S. W. Roberts, A. Harpin, J. Drake, and M. Asghari. Optical dispersion, two-photon absorption and self-phase modulation in silicon waveguides at 1.5μm wavelength. *Appl. Phys. Lett.*, 80(3):416–418, 2002.

[111] T. Tsuchizawa, K. Yamada, H. Fukuda, T. Watanabe, J. Takahashi, M. Takahashi, T. Shoji, E. Tamechika, S. Itabashi, and H. Morita. Microphotonics devices based on silicon microfabrication technology. *IEEE J. Sel. Top. Quantum Electron.*, 11(1):232–240, 2005.

[112] B. Tuck. The chemical polishing of semiconductors. *J. Mater. Sci.*, 10:321–339, 1975.

[113] B. Tuck and A. J. Baker. Chemical etching of {111} and {100} surfaces of InP. *J. Mater. Sci.*, 8:1559–1566, 1973.

[114] C. Vassallo. 1993-1995 optical mode solvers. *Optical and Quantum Electronics*, 29:95–114, 1997.

[115] G. Vijaya Prakash, M. Cazzanelli, Z. Gaburro, L. Pavesi, F. Iacona, G. Franzo, and F. Priolo. Linear and nonlinear optical properties of plasma-enhanced chemical-vapour deposition grown silicon nanocrystals. *J. Mod. Opt.*, 49(5/6):719–730, 2002.

[116] Y. A. Vlasov and S. J. McNab. Losses in single-mode silicon-on-insulator strip waveguides and bends. *Opt. Express*, 12:1622–1631, February 2004.

[117] Y. A. Vlasov, M. O'Bolye, H. F. Hamann, and S. J. McNab. Active control of slow light on a chip with photonic crystal waveguides. *Nature*, 438:65–69, November 2005.

[118] Y. A. Vlasov, F. Xia, L. Sekaric, E. Dulkeith, S. Assefa, W. Green, M. O'Boyle, H. Hamann, and S. J. McNab. Chip-scale all-optical group delay. In *Optical Society of America (OSA) Annual Meeting*, Rochester (NY), USA, October 2006. FThL1.

[119] K. R. Williams and R. S. Muller. Etch rates for micromachining processing. *J. Microelectromech. Syst.*, 5(4):256–268, December 1996.

[120] J. J. Wynne. Optical third-order mixing in GaAs, Ge, Si and InAs. *Phys. Rev.*, 178(3):1295–1301, February 1969.

[121] Q. Xu, V. R. Almeida, R. R. Panepucci, and M. Lipson. Experimental demonstration of guiding and confining light in nanometer-size low-refractive-index material. *Opt. Lett.*, 29(14):1626, 2004.

[122] H. Yamada, M. Shirane, T. Chu, H. Yokoyama, S. Ishida, and Y. Arakawa. Nonlinear-optic silicon-nanowire waveguides. *Jpn. J. Appl. Phys.*, 44(9a):6541–6545, 2005.

[123] K. S. Yee. Numerical solution of initial boundary value problems involving Maxwell's equations in isotropic media. *IEEE Trans. Antennas Propagation*, 14(5):302–307, May 1966.

[124] A. J. Zilkie, J. Meier, P. W. E. Smith, M. Mojahedi, and J. S. Aitchinson. Femtosecond gain and index dynamics in an InAs/InGaAsP quantum dot amplifier operating at $1.55\,\mu$m. *Opt. Express*, 14(23):11453–11459, November 2006.

[125] I. Zubel. The model of etching of (hkl) planes in monocrystalline silicon. *J. Electrochem. Soc.*, 150(6):C391–C400, 2003.

Acknowledgements (German)

Die vorliegende Dissertation entstand während meiner Tätigkeit am Institut für Hochfrequenztechnik und Quantenelektronik (IHQ) der Universität Karlsruhe. Sie war eingebunden in die Projekte A3.1 und A4.4 des Centrums für funktionelle Nanostrukturen (CFN) und in das von der Europäische Union (EU) geförderte Projekt TRIUMPH (grant IST-027638 STP). Es bleibt mir zum Abschluss die erfreuliche Aufgabe, all den Personen zu danken, die im Verlauf der letzten Jahre zum Gelingen dieser Arbeit beigetragen haben.

An erster Stelle danke ich meinen Betreuern Prof. Wolfgang Freude und Prof. Jürg Leuthold für ihr stetiges Interesse an meiner Arbeit, für ihre hilfreichen Anregungen sowie für das Vertrauen und die außergewöhnlich großen Freiräume, die ich genießen durfte.

Meinem Doktorvater Prof. Wolfgang Freude danke ich besonders für viele aufschlussreiche Diskussionen, die zuweilen auch den Rahmen des Wissenschaftlich-Technischen verließen, und sich, in einer von Offenheit geprägten Atmosphäre, Fragestellungen mit philosophischem Hintergrund zuwandten. Die ihm eigene Ruhe und Gelassenheit, die sich auch durch einen zeitweise etwas ungeduldigen Doktoranden nicht erschüttern ließen, brachten mich mehr als einmal auf die nötige gedankliche Distanz und halfen über manche wissenschaftliche Durststrecke hinweg. Dass Prof. Freude im Vorfeld jedweder Abgabetermine zu den unmöglichsten Tages- und Nachtzeiten erreichbar war, kam meinem Arbeitsstil sehr entgegen. Wann er die Zeit gefunden hat, meine Manuskripte mit der gewohnten Gründlichkeit Korrektur zu lesen, ist mir bis heute ein Rätsel.

Prof. Jürg Leuthold, dem Leiter des IHQ, verdanke ich wesentliche inhaltliche Impulse für meine Arbeit. Dank seines fachlichen Überblickes verstand er es hervorragend, eine visionäre Sicht für den Gesamtkontext zu vermitteln, und somit strategische Richtungsentscheidungen zu erleichtern. Dass er so manche Einleitungen meiner Manuskripte nach kritischer Durchsicht verwarf und neu formulierte, wurde durch das jeweilige Ergebnis gerechtfertigt. Als Wissenschaftler strahlte Prof. Leuthold einen unerschütterlichen Optimismus aus, den ich sehr zu schätzen gelernt habe und in dessen Licht plötzlich das Unmögliche möglich erschien.

Prof. Michael Siegel, dem Leiter des Institutes für Mikro- und Nanoelektronische Systeme (IMS), danke ich für Übernahme des Korreferats. Ferner danke ich ihm dafür, dass er mir den Zugang zum Reinraum und Elektronenstrahlschreiber des IMS in denkbar unkomplizierter Weise rund um die Uhr gewährt hat.

Dr. Christopher Poulton danke ich für die Modellierung der Wellenleiterverluste und für die Unterstützung bei zahllosen mathematischen Fragestellungen während der drei Jahre, in denen wir uns ein Büro teilten. Unvergessen sind auch die anregenden wissenschaftlichen Diskussionen, die wir des öfteren, mitleidige Blicke unserer Mitmenschen in

Kauf nehmend, außerhalb des IHQ weiterführten. Dankbar bin ich Chris auch dafür, dass er mich ab und zu im Schach gewinnen ließ.

Desweiteren danke ich Prof. Masafumi Fujii, der uns seinen FDTD-Code zur Verfügung gestellt und uns bei dessen Weiterentwicklung tatkräftig unterstützt hat. Er möge mir nachsehen, dass sich das Fertigstellen mancher Veröffentlichungen trotz seines unermüdlichen Einsatzes etwas länger hingezogen hat.

Dass ich bei meiner Arbeit auf die Einrichtungen zahlreicher anderer Institute an der Universität zurückgreifen konnte, verdanke ich der Hilfsbereitschaft der beteiligten Personen. Dank gebührt den Mitarbeitern des Institutes für Mikro- und Nanoelektronische Systeme (IMS) unter Leitung von Prof. Siegel. Dr. Konstantin Ilin danke ich insbesondere dafür, dass er mich geduldig in die Geheimnisse der Elektronenstrahllithographie eingeführt hat, und dass er bereit war, die verfügbare Maschinenzeit mit mir zu teilen. Dr. Manfred Neuhaus danke ich für die Beratungen bei diversen technologischen Fragestellungen. Alexander Stassen danke ich für das Zersägen von zahllosen Wafern in die unmöglichsten Formen, und Hansjürgen Wermund danke ich für das Aufdampfen diverser Metallschichten.

Dr. Fabian Perez, Jacques Hawecker und Dr. Erich Müller vom Nanostructure Service Laboratory (NSL) des CFN danke ich für einen unkomplizierten Zugang zum Raster-Elektronenmikroskop (REM) und für die Unterstützung bei der Benutzung der Focused-Ion-Beam-Anlage. Mein Dank gilt auch Prof. Dagmar Gerthsen, der Institutsleiterin des Laboratoriums für Elektronenmikroskopie (LEM), an dem derzeit ein Großteil der technischen Infrastruktur des NSL angesiedelt ist. Peter Pfundstein danke ich für die Anfertigung von REM-Bildern, und Winfried Send danke ich für Hilfestellungen bei Fragen der Vakuumtechnik. Dr. Christoph Sürgers vom Physikalischen Institut (PI) danke ich für den Zugang zum Profilometer, der Lackschleuder und der Ionenätzanlage des PI.

Desweiteren danke ich Prof. Uli Lemmer, dem Leiter des Lichttechnischen Institutes (LTI), für den Zugang zum Reinraum. Prof. Thomas Schimmel und Dr. Andreas Pfrang vom Institut für Nanotechnologie danke ich für die Rasterkraftmikroskop-Messungen der Seitenwandrauhigkeit optischer Wellenleiter.

Teile dieser Arbeit sind aus Kooperationen mit externen Partnern hervorgangen. Das Ziel meiner ersten technologischen „Gehversuche" war die Herstellungen von Ringresonatoren in einem passiven InP/InGaAsP-Heterosystem. Für beratende Hilfestellungen möchte ich in diesem Zusammenhang Dr. Dietmar Keiper von der Innovative Processing AG (IPAG) in Duisburg danken. Die Bauteile wurden letztlich am Fraunhofer Institut für Nachrichtentechnik, Heinrich Hertz Institut (HHI) in Berlin prozessiert. Dafür danke ich den Mitarbeitern der Abteilung für Strukturtechnologie unter der Leitung von Dr. Udo Niggebrügge. Besonderer Dank gebührt Ralf Steingrüber, der mich im Kampf gegen meine eigene technologische Unbedarftheit bereitwillig und intensiv unterstützte.

Bei der Prozessierung von Silizium durfte ich auf die Einrichtungen des Institutes für Mikrosystemtechnik (Imtek) an der Universität Freiburg zurückgreifen. Ich danke Prof. Jan Korvink für die Herstellung des Kontaktes, und Prof. Ulrike Wallrabe für die unkomplizierte Handhabung der buchhalterischen Angelegenheiten. Mein Dank gilt auch den Mitarbeitern des Reinraum Service Centers (RSC) unter Leitung von Dr. Michael Wandt. Ich danke insbesondere Armin Baur, der uns bei unseren Ätztests hilfreich und geduldig zur Seite stand, und sich durch nichts aus der Ruhe bringen ließ — auch nicht durch unsere „mikroskopischen" $5 \times 5\,\mathrm{mm}^2$-Chips, die sich gerne auf Nimmerwiedersehen in die Tiefen irgendwelcher Reaktoren oder Ätzbäder absetzten, und die uns schließlich den Beinamen der „Schnipselforscher aus Karlsruhe" einbrachten.

Für die LPCVD-Beschichtung von Silicon-on-insulator-(SOI-)Wafern mit Si_3N_4 danke ich Dr. Arne Albrecht von der Technischen Universität Ilmenau.

Trockenchemische Ätzungen von SOI-Strukturen wurden in Kooperation mit dem Fachgebiet Hochfrequenztechnik/Photonik der Technischen Universität Berlin unter der Leitung von Prof. Klaus Petermann durchgeführt. Ich danke Dr. Lars Zimmermann für die unkomplizierte Zusammenarbeit.

Die im Rahmen dieser Arbeit untersuchten InAs/GaAs-Quantenpunktverstärker wurden uns zur Verfügung gestellt vom Institut für Festkörperphysik der Technischen Universität Berlin, Arbeitsgruppe Prof. Dieter Bimberg. Ich danke in diesem Zusammenhang Dr. Matthias Lämmlin und Christian Meuer.

Die jüngste Generation von SOI-Bauteilen wird derzeit in einer Kooperation mit der Firma ASML und dem European FP6 Network of Excellence ePIXnet hergestellt. Ich danke Günter Janello, Dr. Paul van Dijk und Rudy Pellens für die Unterstützung der Kooperation seitens ASML. Dr. Pieter Dumon vom Interuniversity Microelectronics Center (IMEC) in Leuven (Belgien) danke ich für die Hilfe beim Anfertigen des Layouts.

Für eine großartige Zusammenarbeit danke ich meinen Kollegen Ayan Maitra, Jan Brosi, Jin Wang, Philipp Vorreau, Andrej Marculescu, Dr. Gunnar Böttger, Thomas Vallaitis, Dr. Shunfeng Li, Rene Bonk, Moritz Röger und Dr. Stelios Sygletos, sowie meinen ehemaligen Kollegen Anastasiya Golovina, Dr. Guy-Aymar Chakam, Dr. Richard Ell, Dr. Wolfgang Seitz, Dr. Philipp Wagenblast, Dr. Martin Hugelmann, Dr. Philipp Hugelmann, Lenin Jacome und Oleksiy Shulika.

Jan Brosi danke ich insbesondere für eine fruchtbare Zusammenarbeit in Sachen SOI-Polymermodulator, für wertvolle Tips beim Gebrauch von HFSS, für die kritische Kontrolle meiner Modenkopplungsgleichungen, und darüber hinaus für anregende Gespräche über Fragestellungen jenseits der wissenschaftlich beschreibbaren Welt. Sein Urvertrauen lässt sich allenfalls durch eine Überdosis Kaffee erschüttern, und dann auch nur zeitweise. Philipp Vorreau danke ich für die Unmengen von Messgeräten, die er zu Beginn seiner Arbeit am IHQ größtenteils auf Auktionen zusammengekauft hat, und ohne die ein produktives Arbeiten heute nicht möglich wäre. Ich danke ihm weiterhin für zahlreiche verbale Steilvorlagen, die es uns immer wieder ermöglichten, gemeinsam die Tür zur nahegelegenen Welt des Unsinns aufzubrechen und dem Arbeitsalltag für einige erquickliche Minuten zu entfliehen. Andrej Marculescu danke ich für seine unermüdliche Hilfe bei Computerangelegenheiten sowie beim Beschaffen von Messgeräten. Seine geräumige, strategisch günstig gelegene Wohnung wurde gelegentlich zum Schauplatz abendlicher Veranstaltungen mit Freizeitcharakter. Dass wir uns nach jenem denkwürdigen Abend im Sommer 2006 noch zu seinen Freunden zählen dürfen, rechne ich Andrej hoch an. Gunnar Böttger danke ich für diverse gemeinsame Unternehmungen in der Freizeit sowie für die gewissenhafte photographische Dokumentation derselben. Unvergessen sind auch die unzähligen Muffins und Kuchenstücke, mit denen er uns unsere Kaffeepausen versüßt hat. Thomas Vallaitis danke ich für eine sehr fruchtbare Zusammenarbeit im Labor. Es war beeindruckend, mit welcher Respektlosigkeit und Geduld er so lange an Tsunami und Opal herumjustiert hat, bis endlich die erwünschten 1 300 nm-Impulse herauskamen. Dass die Pump-Probe-Messungen mittlerweile weitgehend automatisiert ablaufen, ist ebenfalls sein Verdienst. Ich hoffe, er kann mir nachsehen, dass mir trotz intensiven Übens der Diphtong in seinem Nachnamen noch immer nicht richtig über die Lippen kommt. Unvergessen sind natürlich auch die gemeinsamen Radtouren. René Bonk danke ich dafür, dass er uns den Rücken freigehalten hat von administrativen Angelegenheiten, die im Zusammenhang mit dem TRIUMPH-Projekt entstehen. Für unzählige Stücke Kuchen danke Moritz Röger bzw. seiner Frau Tina. Dr. Stelios Sygletos, mit dem ich neuerdings das Büro teile, danke ich für die regelmäßige Versorgung mit Lebensmitteln, besonders in der Endphase meiner Promotion.

Desweiteren möchte ich den Hiwis, Studienarbeitern und Diplomanden danken, die mich bei meiner Arbeit unterstützt haben. Es sind dies Jan Brosi, Jin Wang, Felix Glöckler, Leonie Monthé, Lenin Jacome, Michael Waldow, Corinne Tzebia, Boris-Alexander Bolles und Simon Schneider. Dass sich einzelne unter ihnen unter fragwürdigen Vorwänden dem obligatorischen gemeinsamen Mittagessen entzogen, kann ich ihnen dank ihres besonderen wissenschaftlichen Einsatzes nachsehen.

Ich danke weiterhin dem Personal am IHQ für die fortwährende Unterstützung bei technischen, administrativen oder graphischen Problemen. Dagmar Goldmann und Bernadette Lehmann danke ich für die angenehm unbürokratische Handhabung jeglicher administrativer Vorgänge. Seit im Sekretariat Süßigkeiten zum freien Verzehr angeboten werden, kontrolliere ich auch täglich mein Postfach. Die Meinung meines Zahnarztes tut hier nichts zur Sache.

Ilse Kober danke ich für die sorgfältige Anfertigung diverser technischer Zeichungen, für das Einscannen unzähliger Dokumente, und für das Binden vieler Papierstapel, die ich ihr einfach auf den Schreibtisch lege durfte.

Hans Bürger und seinem Team von der Werkstatt, Werner Höhne, Manfred Hirsch sowie den wechselnden Auszubildenden, danke ich für die vielen technisch und handwerklich hervorragenden Spezialanfertigungen, die ich im Laufe der Jahre benötigt habe. Herrn Bürgers Einfallsreichtum war mir dabei eine große Hilfe: Anstelle detaillierter Zeichnungen und Spezifikationen genügte eine kurze Schilderung des „Problems", und ich konnte mich darauf verlassen, das Bauteil in der gewohnten Qualität innerhalb der vereinbarten Zeit zu erhalten. Dafür ein besonderes Dankeschön!

Oswald Speck danke ich für die hervorragende Zusammenarbeit beim Aufbau des Packaging-Labors, aus dessen Tagesgeschäft ich mich mittlerweile weitgehend zurückziehen konnte. Es ist zu großen Teilen sein Verdienst, dass Techniken wie Spleißen, Läppen, Polieren, Spalten, Bonden, das Aufkleben und -Löten von Chips und das Anbringen von Faser-Pigtails mittlerweile zur Selbstverständlichkeit geworden sind.

Werner Podszus danke ich für die Unterstützung bei elektrischen Problemen aller Art sowie für die Instandsetzung zahlreicher Messgeräte, die ich aus diversen Schränken gefischt und mit dem Vermerk „Defekt, ChK, <Datum>" auf seinem Tisch abgestellt habe. Ich danke ihm auch für seine Hilfe bei der Herstellung ungewöhnlicher Lötverbindungen, beispielsweise an Blechblasinstrumenten.

Peter Frey danke ich für die Hilfe beim Auffinden von verschollenen und beim Identifizieren von zufällig aufgefundenen Messgeräten. Besonders für unsere geerbten Freistrahloptiken ist seine akribisch geführte Inventarkartei oft die letzte Hoffnung, nicht aus Mangel aus Information dem Verfall anheimgegeben zu werden. Herrn Frey danke ich außerdem für die tatkräftige Unterstützung bei Fragen der Klima-, Kälte- und Vakuumtechnik, sowie für die hervorragende Lösung der mir persönlich sehr am Herzen liegenden kulinarischen Frage bei diversen Institutsfeiern.

Helena Schneider, René Bonk, Gunnar Böttger, Jan Brosi, Clemens Klöck, Chris Poulton, Thomas Vallaitis und Philipp Vorreau danke ich für die kritische Durchsicht des Manuskriptes. In diesem Zusammenhang geht auch nochmal ein herzliches Dankeschön an Prof. Freude für die unzählige Tricks und Kniffe, mit denen sich das zuweilen exzentrisch anmutende Gebaren von LaTeX in den Griff bekommen ließ.

Meinen Eltern Gertraud und Gunter Koos sowie meiner Schwester Veronika danke ich für die uneingeschränkte Unterstützung, auf die ich mich jederzeit verlassen konnte. Ihnen ist diese Arbeit gewidmet.

List of Publications

Patents

P2 **Koos, C.**; Leuthold, J.; Freude, W.: *Herstellungsverfahren für halbleiterbasierte optische Wellenleiterstrukturen mit speziellen geometrischen Formen*, patent pending, file number DE 10 2007 004 043.3

P1 **Koos, C.**; Leuthold, J.; Freude, W., Brosi, J.-M.: *Elektro-optisches Bauteil mit transparenten Elektroden, kleiner Betriebsspannung und hoher Grenzfrequenz*, patent pending, file number DE 10 2006 045 102.3

Journal Papers

J5 **Koos, C.** ; Jacome, L.; Poulton, C.; Leuthold, J. and Freude, W.: *Nonlinear Silicon-On-Insulator Waveguides for All-Optical Signal Processing*, Opt. Express 15 (2007) 5976–5990

J4 **Koos, C.**; Poulton, C. G.; Zimmermann, L.; Jacome, L.; Leuthold, J.; Freude, W.: *Ideal contour trajectories for single-mode operation of low-loss overmoded waveguides*, IEEE Photon. Technol. Lett. 19 (2007) (819-821)

J3 **Koos, C.**; Fujii, M.; Poulton, C. G.; Steingrueber, R.; Leuthold, J.; Freude, W.: *FDTD-modelling of dispersive nonlinear ring resonators: Accuracy studies and experiments*, IEEE J. Quantum Electron. 42 (2006) 1215–1223

J2 Poulton, C. G.; **Koos, C.**; Fujii, M.; Pfrang, A.; Schimmel, Th.; Leuthold, J.; Freude, W.: *Radiation modes and roughness loss in high index-contrast waveguides*, IEEE J. Sel. Topics Quantum Electron. 12 (2006) 1306–1321

J1 Fujii, M; **Koos, C.**; Poulton, C.; Leuthold, J.; Freude, W.: *Nonlinear FDTD analysis and experimental validation of four-wave mixing in InGaAsP/InP racetrack micro-resonators*, IEEE Photon. Technol. Lett. 18 (2006) 361–363

Book Chapter

B1 Freude, W.; Chakam, G.-A.; Brosi, J.-M.; **Koos, C.**: *Microwave Modelling of Photonic Crystals.* In: Busch, K.; Lölkes, S.; Wehrspohn, R.; Föll, H. (Eds.): *Photonic Crystals – Advances in Design, Fabrication, and Characterization.* Wiley VCH, Berlin 2004, pp.198–214

Conference Contributions

C19 **Koos, C.** ; Jacome, L.; Poulton, C.; Leuthold, J. and Freude, W.: *Nonlinear High Index-Contrast Waveguides with Optimum Geometry,* Nonlinear Photonics (NP), Quebec City, September 2-6, 2007, Paper JWA2

C18 Leuthold, J.; Wang, J.; Vallaitis, T.; **Koos, C.**; Bonk, R.; Marculescu, A.; Vorreau, P.; Sygletos, S.; Freude, W.: *New approaches to perform all-optical signal regeneration,* 9th Intern. Conf. on Transparent Optical Networks (ICTON), July 1–5, 2007, Rome, Italy **(invited)**

C17 Freude, W.; **Koos, C.**; Poulton, C. G.; Fujii, M.; Leuthold, J.: *Minimizing roughness loss for ultracompactly bent high-index contrast waveguides,* 9th Intern. Conf. on Transparent Optical Networks (ICTON), July 1–5, 2007, Rome, Italy **(invited)**

C16 **Koos, C.**; Vallaitis, T.; Bolles, B.-A.; Bonk, R. ; Freude, W.; Laemmlin, M.; Meuer, C.; Bimberg, D.; Ellis, A.; Leuthold, J.: *Gain and phase dynamics in an InAs/GaAs quantum dot amplifier at 1300 nm,* CLEO/Europe-IQEC Conference, Munich, June 17–22, 2007, Paper CI3-1-TUE

C15 **Koos C.**; Poulton C.; Jacome L.; Zimmermann L.; Leuthold J.; Freude W.: *Ideal contour trajectories for low-loss waveguide bends,* ePIXnet, Winterschool, Pontresina, Schweiz, March 11-16, 2007, (**"Best-Poster" Award**)

C14 Freude, W.; Fujii, M.; **Koos, C.**; Brosi, J.; Poulton, C. G.; Wang, J.; Leuthold, J.: *Wavelet FDTD methods and applications in nano-photonics,* 386th WE Heraeus Seminar 'Computational Nano-Photonics', Bad Honnef, February 25–28, 2007 **(invited)**

C13 **Koos, C.**; Poulton, C.; Jacome, L.; Zimmermann, L.; Leuthold, J.; Freude, W.: *Ideal trajectory for ultracompact low-loss waveguide bends,* 32th European Conf. Opt. Commun. (ECOC'06), Cannes, September 24–28, 2006, Paper Tu1.4.6

C12 Freude, W.; Maitra, A.; Wang, J.; **Koos, C.**; Poulton, C.; Fujii, M.; Leuthold, J.: *All-optical signal processing with nonlinear resonant devices,* Proc. 8th Intern. Conf. on Transparent Optical Networks (ICTON'06), June 18–22, 2006, Nottingham, UK. Paper We.D2.1, vol. 2, pp. 215–219 **(invited)**

C11 Fujii, M.; **Koos, C.**; Poulton, C.; Leuthold, J.; Freude, W.: *FDTD modelling and experimental verification of FWM in semiconductor micro-resonators,* Proc. 15th

International Workshop on Optical Waveguide Theory and Numerical Modelling (OWTNM'06), Varese, Italy, April 20–21, 2006, p. 79

C10 Freude, W.; Poulton, C. G.; **Koos, C.**; Brosi, J.; Fujii, M.; Pfrang, A.; Schimmel, Th.; Müller, M.; Leuthold, J.: *Nanostrip and photonic crystal waveguides: The modelling of imperfections*, Proc. 15th International Workshop on Optical Waveguide Theory and Numerical Modelling (OWTNM'06), Varese, Italy, April 20–21, 2006, pp. 1–2 **(invited)**

C9 Poulton, C.; **Koos, C.**; Fujii, M.; Pfrang, A.; Schimmel, Th.; Leuthold, J.; Freude, W.: *Roughness in high index-contrast waveguides*, European Cooperation in the Field of Scientific and Technical Research (COST), COST P11 Workshop and WG Meeting, Twente; October 2–4, 2005

C8 Freude, W.; Poulton, C.; **Koos, C.**; Brosi, J.; Fujii, M.; Müller, M.; Leuthold, J.; Pfrang, A.; Schimmel, Th.: *Nano-optical waveguide devices with random deformations and nonlinearities*, Conf. Dig. 3rd Intern. Conf. on Materials for Advanced Technologies (ICMAT'05), Symposium F: Nano-Optics and Microsystems, Singapore, July 3–8, 2005, Paper F-2-IN3 **(invited)**

C7 Freude, W.; Poulton, C.; Brosi, J.; **Koos, C.**; Glöckler, F.; Wang, J.; Fujii, M.: *Integrated optical waveguide devices: Design, modeling and fabrication*, Proc. 16th Asia Pacific Microwave Conference (APMC'04), New Delhi, India, December 15–18, 2004 **(invited)**

C6 Poulton, C. G.; **Koos, C.**; Müller, M.; Glöckler, F.; Wang, J.; Fujii, M.; Leuthold, J.; Freude, W.: *Sidewall roughness and deformations in high index-contrast waveguides and photonic crystals*, Proc. 17th Annual Meeting of the IEEE Lasers and Electro-Optics Society (LEOS 2004), Puerto Rico, USA, November 7–11, 2004 vol. 2 pp. 949–950

C5 Freude, W.; Brosi, J.; Glöckler, F.; **Koos, C.**; Poulton, C.; Wang, J.; Fujii, M.: *High-index optical waveguiding structures*, Conf. Proc. Optical Society of America Annual Meeting (OSA 2004), Rochester (NY), USA, October 10–14, 2004, paper TuS1 **(invited)**

C4 Poulton, C. G.; **Koos, C.**; Freude, W.; Fujii, M.: *Attenuation of optical strip waveguides with rough sidewalls*, Proc. 29th Intern. Conf. on Infrared and Millimeter Waves and 12th International Conference on Terahertz Electronics (IR-MMW/THz'04), Karlsruhe, Germany, 27.09.–01.10.2004 pp. 141–142

C3 Freude, W.; Poulton, C.; **Koos, C.**; Brosi, J.; Glöckler, F.; Wang, J.; Chakam, G.-A.; Fujii, M.: *Design and fabrication of nanophotonic devices*, Proc. 6th Intern. Conf. on Transparent Optical Networks (ICTON'04), July 4–8, 2004, Wroclaw, Poland. Mo.A.4 vol. 1 pp. 4–9 **(invited)**

C2 Freude, W.; Brosi, J.-M.; Chakam, G.-A.; Glöckler, F.; **Koos, C.**; Poulton, C.; Wang, J.: *Modelling and design of nanophotonic devices*, URSI Kleinheubacher Tagung, Miltenberg, Germany; September/October 2003 **(invited)**

C1 Poulton, C.; Brosi, J.; Glöckler, F.; **Koos, C.**; Wang, J.; Freude, W.; Ilin, K.; Siegel M.; Fujii, M.: *Modelling, design and realisation of SOI-based nanophotonic devices*; Conference on Functional Nanostructures 2003 (CFN03), Karlsruhe, Germany. September/October 2003

Curriculum Vitae

Christian Koos

born on April 13th, 1978
in Heilbronn, Germany
Citizenship: German

Education

08/02–07/07 PhD studies at the Institute of High-Frequency and Quantum Electronics (IHQ), University of Karlsruhe

08/01–07/02 Visiting Scientist at Massachusetts Institute of Technology (MIT), Ultrafast Optics Group, Research Laboratory of Electronics (RLE), Department of Electrical Engineering and Computer Science

10/97–07/02 Studies at University of Karlsruhe, Germany
Major: Electrical Engineering
Specialisation: Optical Communications
Degree: Dipl.-Ing.

08/88–06/97 Adolf-Schmitthenner-Gymnasium Neckarbischofsheim (secondary school)

Internships and Work Experience

03/01–06/01 Internship at Nortel Networks Optical Components (Switzerland) AG, Zürich, Switzerland
Project: Measurement of spectral gain in semiconductor lasers

10/00–02/01 Teaching assistant for Semiconductor Devices (Institute of High-Frequency and Quantum Electronics, University of Karlsruhe)

06/00–01/01 Student assistant at the Institut für Höchstfrequenztechnik und Elektronik (IHE), University of Karlsruhe

01/00–03/00 Teaching assistant for Electrophysics (Lichttechnisches Institut, University of Karlsruhe)

10/99–04/00 Teaching assistant for Probability Theory (Institut für Nachrichtentechnik, University of Karlsruhe)

10/98–09/99 Teaching assistant for Mathematics (Mathematisches Institut, University of Karlsruhe)

02/98–03/98 Internship at Robert Bosch GmbH, Stuttgart-Feuerbach, Germany Wiring of automation systems

08/97–10/97 Internship at a steel-construction company and a company for automation systems, practical training (mechanical workshop)

Honors and Awards

03/2000 *"Vordiploms-Preis des Instituts für Prozeßmeßtechnik und Prozeßleittechnik"* (Prediploma award)

06/1997 *"Adolf-Schmitthenner-Preis der Stadt Neckarbischofsheim"* for best Abitur (diploma from German secondary school qualifying for university admission or matriculation)

03/1997 2nd prize at the *"39. Schülerwettbewerb des Landtages von Baden-Württemberg zur Förderung der politischen Bildung"* (student-competition, organized by the parliament of Baden-Württemberg)

Scholarships

03/2002 Presidential Fellowship for PhD studies at the Massachusetts Institute of Technology (declined in favour of a position in Germany)

08/01–08/02 Dr.-Jürgen-Ulderup-Stiftung

08/01–06/02 German-American Fulbright-Kommission

04/00–12/02 DaimlerChrysler Stipendiatenprogramm Forschung und Technologie

12/97–07/02 Studienstiftung des deutschen Volkes (German Scholarship Foundation)